How to Harness Renewable Energy - Solar, Wind, and Geothermal Power Generation

A Comprehensive Guide to Sustainable Energy Solutions: Implementing Solar Panels, Wind Turbines, and Geothermal Systems for Homes and Businesses

Jack Homer

Copyright © 2024 by Jack Homer

All rights reserved. No part of this publication may be reproduced, distributed, or transmitted in any form or by any means, including photocopying, recording, or other electronic or mechanical methods, without the prior written permission of the author, except in the case of brief quotations embodied in critical reviews and certain other noncommercial uses permitted by copyright law.

Table of Contents

Introduction
 - The importance of proper septic system management
 - Who this book is for

Part I: Understanding Renewable Energy Sources
Chapter 1: The Basics of Solar Energy
 - How Solar Panels Work
 - Types of Solar Panels
 - Advantages and Disadvantages of Solar Energy

Chapter 2: The Basics of Wind Energy
 - How Wind Turbines Work
 - Types of Wind Turbines
 - Advantages and Disadvantages of Wind Energy

Chapter 3: The Basics of Geothermal Energy
 - How Geothermal Systems Work
 - Types of Geothermal Systems
 - Advantages and Disadvantages of Geothermal Energy

Part II: Implementing Renewable Energy Solutions for Homes
Chapter 4: Assessing Your Home's Energy Needs
 - Conducting an Energy Audit
 - Determining Your Renewable Energy Potential

Chapter 5: Designing a Solar Panel System for Your Home
 - Sizing Your Solar Panel Array
 - Choosing the Right Components
 - Installation and Maintenance

Chapter 6: Designing a Wind Turbine System for Your Home
- Assessing Wind Resources in Your Area
- Choosing the Right Wind Turbine
- Installation and Maintenance

Chapter 7: Designing a Geothermal System for Your Home
- Evaluating Your Home's Geothermal Potential
- Choosing the Right Geothermal System
- Installation and Maintenance

Part III: Implementing Renewable Energy Solutions for Businesses

Chapter 8. Assessing Your Business's Energy Needs
- Conducting a Commercial Energy Audit
- Determining Your Renewable Energy Potential

Chapter 9: Designing a Solar Panel System for Your Business
- Sizing Your Commercial Solar Panel Array
- Choosing the Right Components
- Installation, Maintenance, and Monitoring

Chapter 10: Designing a Wind Turbine System for Your Business
- Assessing Wind Resources at Your Business Location
- Choosing the Right Commercial Wind Turbine
- Installation, Maintenance, and Monitoring

Chapter 11: Designing a Geothermal System for Your Business
- Evaluating Your Business's Geothermal Potential
- Choosing the Right Commercial Geothermal System
- Installation, Maintenance, and Monitoring

Part IV: Financing and Incentives for Renewable Energy
Chapter 12: Government Incentives and Tax Credits
 - Financing Options for Renewable Energy Projects
 - Return on Investment and Long-Term Savings

Introduction

Imagine a world where you could power your home or business without relying on the grid, while significantly reducing your carbon footprint and energy bills. This reality is closer than you might think. As climate change becomes an increasingly pressing issue, and traditional energy sources continue to deplete, the need for sustainable energy solutions has never been greater.

Did you know that by harnessing the power of the sun, wind, and Earth, you can not only reduce your dependence on fossil fuels but also save thousands of dollars in the long run? Many homeowners and businesses are already making the switch, but navigating the world of renewable energy can be daunting. That's where "How to Harness Renewable Energy - Solar, Wind, and Geothermal Power Generation" comes in.

In this comprehensive guide, you'll discover the secrets to implementing sustainable energy solutions tailored to your unique needs. Whether you're a homeowner looking to slash your energy bills or a business owner aiming to minimize your environmental impact, this book will empower you with the knowledge and tools to make informed decisions.

Inside, you'll find step-by-step instructions on assessing your energy needs, designing and installing solar panels, wind turbines, and geothermal systems, and optimizing their performance for maximum efficiency. You'll also learn about the latest advancements in renewable energy technology, as well as the financial incentives and financing options available to make your transition to clean energy more affordable.

By the end of this book, you'll have the confidence and expertise to embark on your own renewable energy journey. You'll not only be doing your part to combat climate change but also enjoying the benefits of lower energy bills, increased energy independence, and a more sustainable lifestyle.

Don't wait another moment to take control of your energy future. With "How to Harness Renewable Energy - Solar, Wind, and Geothermal Power Generation: A Comprehensive Guide to Sustainable Energy Solutions," you'll have the power to create a cleaner, greener world, one home and one business at a time. Start your journey today and discover the limitless potential of renewable energy!

Part I: Understanding Septic Systems
Chapter 1
The Basics of Solar Energy
How Solar Panels Work

Solar energy is one of the most abundant and accessible forms of renewable energy available today. By harnessing the power of the sun, you can generate clean, sustainable electricity for your home or business. In this chapter, we'll dive into the fundamentals of solar energy and explore how solar panels work to convert sunlight into usable energy.

How Solar Panels Work

Imagine you're standing outside on a sunny day, feeling the warmth of the sun on your skin. That warmth is actually energy in the form of photons, which are tiny particles of light. Solar panels are designed to capture these photons and convert them into electricity through a process called the photovoltaic effect. Here's a step-by-step breakdown of how solar panels work:

1. Photovoltaic Cells: Solar panels are made up of many smaller units called photovoltaic cells. These cells are typically made from silicon, a semiconductor material that is sensitive to light.

2. Photon Absorption: When sunlight hits a photovoltaic cell, the photons are absorbed by the silicon material. This absorption causes electrons within the silicon atoms to become energized and break free from their orbits.

3. Electron Flow: The photovoltaic cell is designed with two layers of silicon, one positively charged (p-type) and one negatively charged (n-type). As the energized electrons break free, they naturally flow from the n-type layer to the p-type layer, creating an electrical current.

4. Electrical Circuit: The flow of electrons is captured by metal conductors placed on the top and bottom of the photovoltaic cell. These conductors are connected to wires that feed into an inverter.

5. DC to AC Conversion: The electrical current generated by the photovoltaic cells is direct current (DC). However, most homes and businesses use alternating current (AC). The inverter's job is to convert the DC electricity into AC electricity, making it suitable for use in your electrical system.

6. Powering Your Home or Business: Once the electricity is converted to AC, it flows from the inverter to your electrical panel, where it can be distributed to power your lights, appliances, and other electrical devices. Any excess electricity can be fed back into the grid or stored in batteries for later use.

To better understand the scale of a solar panel system, let's consider an example. Imagine you have a 5-kilowatt (kW) solar panel system installed on your roof. This system might consist of 20 individual solar panels, each rated at 250 watts. On a sunny day, these panels can collectively generate 5 kilowatt-hours (kWh) of electricity per hour, which translates to approximately 30-40 kWh of electricity per day (depending on your location and weather conditions).

Now that you have a better understanding of how solar panels work, you can appreciate the incredible potential of solar energy. By installing a solar panel system, you can harness the power of the sun to generate clean, renewable electricity for your home or business, reducing your reliance on fossil fuels and lowering your energy bills in the process.

In the next section, we'll explore the different types of solar panels available and help you determine which option might be best suited for your specific needs and goals.

Types of Solar Panels

When it comes to installing solar panels, you have several options to choose from. Each type of solar panel has its own unique characteristics, advantages, and disadvantages. In this section, we'll explore the three main types of solar panels available on the market today: monocrystalline, polycrystalline, and thin-film.

1. Monocrystalline Solar Panels

Monocrystalline solar panels are the most efficient and durable option available. They are made from a single, high-quality silicon crystal, which gives them their distinctive black color and rounded edges. Here are some key features of monocrystalline panels:

- Efficiency: Monocrystalline panels typically have the highest efficiency ratings, ranging from 17% to 22%. This means they can generate more electricity from the same amount of sunlight compared to other types of panels.
- Durability: These panels are highly resilient and can withstand harsh weather conditions, making them a good choice for areas with strong winds, heavy snow, or high temperatures.
- Aesthetics: The sleek, black appearance of monocrystalline panels is often considered more visually appealing, making them a popular choice for homeowners who prioritize aesthetics.
- Cost: Due to their high efficiency and manufacturing process, monocrystalline panels tend to be more expensive than other options.

Imagine you have limited roof space but high energy needs. In this case, monocrystalline panels might be the best choice for you, as they can generate the most electricity from a smaller surface area.

2. Polycrystalline Solar Panels

Polycrystalline solar panels are made from multiple silicon crystals melted together and cut into wafers. They have a blue, speckled appearance and are slightly less efficient than monocrystalline panels. Here are some key features of polycrystalline panels:

- Efficiency: Polycrystalline panels have efficiency ratings ranging from 15% to 17%, which is slightly lower than monocrystalline panels but still suitable for most applications.
- Cost: Due to their simpler manufacturing process, polycrystalline panels are generally less expensive than monocrystalline panels, making them a more budget-friendly option.
- Durability: While not as durable as monocrystalline panels, polycrystalline panels are still reliable and can withstand most weather conditions.
- Aesthetics: The blue, speckled appearance of polycrystalline panels may be less appealing to some homeowners who prioritize a sleek, uniform look.

Consider a scenario where you have ample roof space and want to maximize your solar energy output while keeping costs down. In this case, polycrystalline panels could be a good choice, as they offer a balance between efficiency and affordability.

3. Thin-Film Solar Panels

Thin-film solar panels are made by depositing thin layers of photovoltaic material onto a substrate, such as glass, plastic, or metal. They are the least efficient option but offer unique advantages in certain applications. Here are some key features of thin-film panels:

- Efficiency: Thin-film panels have the lowest efficiency ratings, typically ranging from 7% to 13%. This means they require more surface area to generate the same amount of electricity as crystalline panels.
- Flexibility: Some thin-film panels are designed to be flexible, making them suitable for applications where traditional rigid panels may not be practical, such as on curved surfaces or portable devices.
- Weight: Thin-film panels are generally lighter than crystalline panels, which can be advantageous for certain installations or transportation.
- Cost: While thin-film panels are less efficient, they can be less expensive to manufacture, making them a cost-effective option for large-scale installations or projects with ample space.

Imagine you're planning to install solar panels on a large commercial building with a flat roof. In this case, thin-film panels might be a suitable choice, as they can cover a larger area at a lower cost, even though they are less efficient than crystalline panels.

When deciding which type of solar panel to install, consider factors such as your available space, energy needs, budget, and aesthetic preferences. Keep in mind that the efficiency and performance of solar panels can also be affected by factors like temperature, shading, and the angle at which they are installed.

In the next section, we'll explore the advantages and disadvantages of solar energy, helping you weigh the pros and cons of investing in a solar panel system for your home or business.

ADVANTAGES AND DISADVANTAGES OF SOLAR ENERGY

Before deciding to invest in a solar panel system, it's essential to understand the advantages and disadvantages of solar energy. In this section, we'll explore the pros and cons to help you make an informed decision.

Advantages of Solar Energy

1. Clean and Renewable: Solar energy is a clean, renewable resource that does not produce greenhouse gases or other pollutants. By harnessing the power of the sun, you can reduce your carbon footprint and contribute to a more sustainable future.

2. Lower Energy Bills: By generating your own electricity, you can significantly reduce or even eliminate your reliance on the grid, leading to substantial savings on your energy bills over time.

3. Tax Incentives and Rebates: Many governments offer tax incentives, rebates, and grants to encourage the adoption of solar energy. These financial benefits can help offset the initial cost of installing a solar panel system.

4. Low Maintenance: Solar panels have no moving parts and require minimal maintenance. Most systems come with a warranty of 25 years or more, ensuring long-term reliability and performance.

5. Energy Independence: By generating your own electricity, you can become less dependent on the grid and protect yourself from future energy price hikes.

6. Versatility: Solar panels can be installed on a variety of surfaces, including roofs, walls, and even the ground. They can be used to power homes, businesses, and even remote locations that are not connected to the grid.

Imagine you live in an area with high electricity rates and frequent power outages. By installing a solar panel system with battery storage, you can generate your own clean electricity, reduce your energy bills, and maintain power during grid failures.

Disadvantages of Solar Energy

1. High Upfront Costs: The initial cost of installing a solar panel system can be substantial, even with tax incentives and rebates. However, the long-term savings on energy bills can often justify the investment.

2. Weather Dependent: Solar panels rely on sunlight to generate electricity, which means their performance can be affected by weather conditions. Cloudy days, shorter winter days, and seasonal variations can impact the amount of electricity generated.

3. Energy Storage Costs: If you want to store excess solar energy for use during the night or on cloudy days, you'll need to invest in a battery storage system, which can add to the overall cost of your solar installation.

4. Space Requirements: Solar panels require a significant amount of space to generate enough electricity to power a home or business. If you have limited roof space or land, you may not be able to install a large enough system to meet your energy needs.

5. Potential Roof Issues: Installing solar panels on your roof can be complex, and not all roofs are suitable for installation. Some roofing materials, such as slate or cedar shingles, may make installation more challenging or expensive.

6. Recycling Concerns: As solar panels reach the end of their lifespan, there are concerns about the recycling and disposal of the materials used in their construction, such as heavy metals and rare earth elements.

Consider a scenario where you live in an area with frequent cloudy days and limited roof space. In this case, a solar panel system may not be the most practical or cost-effective solution for your energy needs.

While the advantages of solar energy are significant, it's crucial to weigh them against the potential drawbacks and consider your specific circumstances. By understanding both the pros and cons, you can make an informed decision about whether solar energy is the right choice for your home or business.

In the next chapter, we'll explore the basics of wind energy and how wind turbines can be used to generate clean, renewable electricity.

Chapter 2
The Basics of Wind Energy
How Wind Turbines Work

Wind energy is another clean, renewable resource that has gained significant popularity in recent years. By harnessing the power of the wind, we can generate electricity without producing greenhouse gases or other pollutants. In this chapter, we'll explore the fundamentals of wind energy and how wind turbines work to convert the kinetic energy of the wind into usable electricity.

How Wind Turbines Work
Have you ever stood outside on a windy day and felt the force of the wind pushing against you? That force is actually kinetic energy, which can be captured and converted into electricity using wind turbines. Here's a step-by-step breakdown of how wind turbines work:

1. Wind Flow: As wind blows across the land or sea, it creates a flow of kinetic energy. The strength and consistency of this wind flow are key factors in determining the potential for wind energy generation in a given location.

2. Blade Rotation: Wind turbines have large blades, typically made from fiberglass or carbon fiber, that are designed to catch the wind. As the wind blows, it exerts a force on the blades, causing them to rotate. The shape and angle of the blades are carefully engineered to maximize the amount of energy captured from the wind.

3. Shaft and Gearbox: The rotating blades are connected to a shaft, which spins as the blades turn. In most wind turbines, the shaft is connected to a gearbox, which increases the rotational speed of the shaft to a level suitable for electricity generation.

4. Generator: The high-speed shaft from the gearbox is connected to a generator, which converts the mechanical energy of the spinning shaft into electrical energy. As the shaft spins, it causes the generator's rotor to rotate within a magnetic field, inducing an electrical current in the generator's windings.

5. Voltage Transformation: The electricity generated by the wind turbine is typically at a lower voltage than what is required for distribution. A transformer located at the base of the wind turbine or in a nearby substation steps up the voltage to prepare the electricity for transmission.

6. Grid Connection: The high-voltage electricity is then fed into the electrical grid, where it can be distributed to homes, businesses, and other consumers. Any excess electricity can be stored in batteries or other energy storage systems for later use.

To better understand the potential of wind energy, let's consider an example. Imagine a large, onshore wind turbine with a capacity of 3 megawatts (MW). On a day with strong, consistent winds, this single turbine could generate enough electricity to power hundreds of homes. Now, picture a wind farm with dozens or even hundreds of these turbines, and you can begin to grasp the immense potential of wind energy.

It's important to note that the efficiency and performance of wind turbines can be influenced by several factors, including wind speed, wind consistency, blade design, and the height of the turbine. To maximize energy generation, wind turbines are often installed in areas with strong, steady winds, such as hilltops, plains, or offshore locations.

As you can see, wind turbines are a marvel of engineering, designed to harness the power of the wind and convert it into clean, renewable electricity. By understanding how these impressive machines work, you can better appreciate the role wind energy can play in creating a more sustainable future.

In the next section, we'll explore the different types of wind turbines available and help you understand which options might be best suited for various applications, from small-scale residential installations to large, utility-scale wind farms.

Types of Wind Turbines

Wind turbines come in various shapes, sizes, and designs, each tailored to specific applications and environments. In this section, we'll explore the main types of wind turbines, including horizontal-axis wind turbines (HAWTs), vertical-axis wind turbines (VAWTs), and small-scale wind turbines for residential or off-grid use.

1. Horizontal-Axis Wind Turbines (HAWTs)

HAWTs are the most common type of wind turbine and are typically used in large-scale wind farms. They have a propeller-like design, with blades that rotate around a horizontal axis. Here are some key features of HAWTs:

- Efficiency: HAWTs are highly efficient, with some models capable of converting up to 50% of the wind's kinetic energy into electricity. Their efficiency is due to their design, which allows the blades to always face the wind and capture the maximum amount of energy.
- Size: HAWTs can be incredibly large, with some offshore turbines reaching heights of over 200 meters (656 feet) and blade lengths exceeding 80 meters (262 feet). These massive turbines can have capacities of up to 12 MW or more, making them well-suited for utility-scale energy generation.
- Wind Direction: To maximize energy capture, HAWTs need to face the wind directly. They are equipped with a yaw drive system that rotates the turbine to align with the prevailing wind direction.

- Noise: Due to their size and the speed at which the blades rotate, HAWTs can generate noise, which may be a concern for nearby residents. However, advancements in technology and careful site selection can help mitigate noise issues.

Imagine a large offshore wind farm with dozens of HAWTs, each standing tall above the waves. These impressive structures can harness the strong, consistent winds found at sea, generating vast amounts of clean electricity for coastal communities.

2. Vertical-Axis Wind Turbines (VAWTs)
VAWTs have blades that rotate around a vertical axis, perpendicular to the ground. They are less common than HAWTs but offer some unique advantages in certain applications. Here are some key features of VAWTs:

- Omnidirectional: VAWTs can capture wind from any direction without the need for a yaw drive system, making them well-suited for areas with variable or turbulent wind conditions.
- Compact: VAWTs have a smaller footprint than HAWTs, making them a good choice for urban or space-constrained environments.
- Low Noise: Due to their smaller size and lower blade speeds, VAWTs generally produce less noise than HAWTs, which can be advantageous in residential or urban settings.
- Maintenance: The gearbox and generator of a VAWT can be placed near the ground, making maintenance and repairs easier and less costly compared to HAWTs.

- Efficiency: While some VAWTs can be less efficient than HAWTs, advancements in design and materials have led to improved performance in recent years.

Picture a small, urban community with limited open space. By installing VAWTs on rooftops or in public spaces, this community can generate clean electricity without the need for large, open areas required by HAWTs.

3. Small-Scale Wind Turbines

Small-scale wind turbines, also known as microwind turbines, are designed for residential, off-grid, or remote applications. They are much smaller than utility-scale turbines and can be used to power homes, farms, or small businesses. Here are some key features of small-scale wind turbines:

- Size: Small-scale wind turbines typically have capacities ranging from a few hundred watts to several kilowatts. They have much smaller blades and towers compared to large-scale turbines.
- Application: These turbines are well-suited for off-grid or remote locations where access to the electrical grid is limited or unavailable. They can be used in combination with solar panels and batteries to create a reliable, renewable energy system.
- Installation: Small-scale wind turbines can be installed on rooftops, poles, or towers, depending on the specific model and application.

- Cost: While the cost per kilowatt-hour of electricity generated by small-scale wind turbines may be higher than utility-scale turbines, they can still provide significant savings compared to traditional energy sources in off-grid or remote settings.

Imagine a remote cabin nestled in the mountains, far from the nearest power lines. By installing a small-scale wind turbine and combining it with solar panels and batteries, the cabin's residents can enjoy clean, reliable electricity without the need for a grid connection.

As you can see, there are various types of wind turbines designed to suit different needs and environments. Whether you're considering a large-scale wind farm or a small, off-grid installation, understanding the different types of wind turbines available can help you make an informed decision.

In the next section, we'll explore the advantages and disadvantages of wind energy, providing a balanced perspective on this clean, renewable energy source.

Advantages and Disadvantages of Wind Energy

Wind energy offers a promising alternative to traditional fossil fuels, but it's essential to understand both the advantages and disadvantages of this renewable energy source. In this section, we'll explore the pros and cons of wind energy to help you make an informed decision about its potential role in your energy mix.

Advantages of Wind Energy

1. Clean and Renewable: Wind energy is a clean, renewable resource that does not produce greenhouse gases or other pollutants during operation. By harnessing the power of the wind, we can generate electricity without contributing to climate change or air pollution.

2. Abundant and Widespread: Wind is a widely available resource, with many areas around the world experiencing sufficient wind speeds to generate electricity. This abundance makes wind energy a viable option for many regions, both onshore and offshore.

3. Low Operational Costs: Once a wind turbine is installed, the operational costs are relatively low. The fuel (wind) is free, and maintenance costs are generally lower compared to traditional power plants.

4. Economic Benefits: The wind energy industry creates jobs in manufacturing, installation, maintenance, and support services. It can also provide economic benefits to local communities through land lease payments and increased tax revenue.

5. Rapid Deployment: Wind turbines can be installed relatively quickly compared to traditional power plants, allowing for a faster transition to renewable energy sources.

6. Complementary to Other Renewables: Wind energy can be combined with other renewable energy sources, such as solar power, to create a more reliable and stable energy mix.

Imagine a rural community that has historically relied on coal-fired power plants. By transitioning to wind energy, this community can enjoy cleaner air, create new jobs, and contribute to the fight against climate change.

Disadvantages of Wind Energy

1. Intermittency: Wind is not a constant resource, and wind speeds can vary significantly over time. This intermittency can make it challenging to rely solely on wind energy for electricity generation and may require the use of energy storage systems or backup power sources.

2. Noise and Visual Impact: Wind turbines can generate noise and have a visual impact on the landscape. This can be a concern for nearby residents and may lead to opposition to wind farm development in some areas.

3. Wildlife Impacts: Wind turbines can pose a threat to birds and bats, causing collisions and fatalities. However, proper siting and the use of modern, wildlife-friendly turbine designs can help mitigate these impacts.

4. Transmission Infrastructure: Wind farms are often located in remote areas, far from population centers. This can require the construction of new transmission lines to deliver the electricity to consumers, adding to the overall cost and complexity of wind energy projects.

5. Land Use: Large-scale wind farms can require significant amounts of land, which may compete with other land uses such as agriculture or natural habitats. However, wind turbines can coexist with other land uses, such as farming or grazing.

6. Rare Earth Elements: Some wind turbines use rare earth elements in their magnets, which can be subject to supply chain disruptions and environmental concerns associated with mining and processing.

Consider a scenario where a proposed offshore wind farm faces opposition due to concerns about its impact on coastal views and marine life. In this case, developers would need to work closely with stakeholders to address these concerns and find a balanced solution.

While wind energy offers significant advantages, it's crucial to weigh them against the potential drawbacks and consider the specific circumstances of each project. By understanding both the pros and cons, policymakers, developers, and individuals can make informed decisions about the role of wind energy in the transition to a clean energy future.

Chapter 3
The Basics of Geothermal Energy
How Geothermal Systems Work

Geothermal energy is a renewable energy source that harnesses the Earth's internal heat to generate electricity and provide heating and cooling for buildings. Unlike wind and solar energy, which are intermittent and depend on weather conditions, geothermal energy is a reliable, baseload power source that can operate around the clock. In this chapter, we'll explore the fundamentals of geothermal energy and how geothermal systems work to provide clean, renewable energy.

How Geothermal Systems Work
The Earth's core is incredibly hot, with temperatures reaching up to 6,000°C (10,800°F). This heat radiates outward, warming the rock and water in the Earth's crust. Geothermal systems tap into this naturally occurring heat to generate electricity and provide heating and cooling. Here's a step-by-step breakdown of how geothermal systems work:

1. Resource Identification: Geothermal resources are typically found in areas with high heat flow, such as near tectonic plate boundaries, volcanic regions, or deep underground. Geologists and engineers use various techniques, such as seismic surveys and temperature gradient measurements, to identify potential geothermal resources.

2. Well Drilling: Once a suitable geothermal resource is identified, wells are drilled to access the hot water or steam. These wells can be several kilometers deep, depending on the location and characteristics of the geothermal reservoir.

3. Fluid Extraction: Hot water or steam is pumped from the geothermal reservoir to the surface through the production well. In some cases, the geothermal fluid may be a mixture of water and steam, which can be separated at the surface.

4. Electricity Generation: For electricity production, the hot water or steam is directed to a geothermal power plant. In a flash steam plant, the high-pressure fluid is rapidly vaporized, and the resulting steam drives a turbine connected to a generator, producing electricity. In a binary cycle plant, the geothermal fluid heats a secondary fluid with a lower boiling point, which then vaporizes and drives the turbine.

5. Direct Use Applications: Geothermal energy can also be used directly for heating and cooling buildings, greenhouses, and industrial processes. In these applications, the hot water is circulated through heat exchangers to transfer its thermal energy to the target space or process.

6. Fluid Reinjection: After the geothermal fluid has been used for electricity generation or direct use applications, it is typically reinjected back into the reservoir through injection wells. This helps to maintain pressure in the reservoir and prolong the life of the geothermal system.

To better understand the potential of geothermal energy, let's consider an example. Imagine a geothermal power plant located in a volcanic region. This plant could generate enough electricity to power thousands of homes and businesses, providing a reliable, clean energy source that operates continuously, regardless of weather conditions.

It's important to note that the efficiency and performance of geothermal systems can be influenced by factors such as the temperature and flow rate of the geothermal fluid, the depth and permeability of the reservoir, and the design of the power plant or direct use system.

As you can see, geothermal systems are a powerful and versatile renewable energy technology, capable of providing both electricity and thermal energy. By understanding how these systems work, you can better appreciate the role geothermal energy can play in creating a more sustainable future.

In the next section, we'll explore the different types of geothermal systems, including conventional hydrothermal systems, enhanced geothermal systems (EGS), and geothermal heat pumps, and discuss their applications and potential.

Types of Geothermal Systems

Geothermal energy systems come in various forms, each with its own unique characteristics, applications, and potential. In this section, we'll explore the main types of geothermal systems, including conventional hydrothermal systems, enhanced geothermal systems (EGS), and geothermal heat pumps.

1. Conventional Hydrothermal Systems

Conventional hydrothermal systems are the most common type of geothermal energy system and have been used for electricity generation and direct use applications for decades. These systems rely on naturally occurring underground reservoirs of hot water or steam. Here are some key features of conventional hydrothermal systems:

- Resource Characteristics: Hydrothermal reservoirs are typically found in areas with high heat flow, such as near tectonic plate boundaries or volcanic regions. The geothermal fluid in these reservoirs can range from hot water to steam, with temperatures often exceeding 150°C (300°F).
- Electricity Generation: High-temperature hydrothermal systems (>150°C) are well-suited for electricity generation. The hot water or steam is extracted from the reservoir and used to drive turbines in flash steam or binary cycle power plants.

- Direct Use Applications: Lower-temperature hydrothermal systems (<150°C) are often used for direct use applications, such as space heating, greenhouse heating, and industrial processes. The hot water is circulated through heat exchangers to transfer its thermal energy to the target application.
- Sustainability: Hydrothermal systems can be sustainable if managed properly, with fluid reinjection helping to maintain reservoir pressure and prolong the life of the geothermal field.

Imagine a city located near a hydrothermal reservoir. This city could use the geothermal resource to generate electricity for homes and businesses, as well as to heat buildings and greenhouses, reducing its reliance on fossil fuels and contributing to a cleaner environment.

2. Enhanced Geothermal Systems (EGS)

Enhanced Geothermal Systems, also known as engineered geothermal systems, are a promising technology that aims to expand the potential of geothermal energy by creating artificial geothermal reservoirs in hot, dry rock formations. Here are some key features of EGS:

- Resource Characteristics: EGS targets deep, hot rock formations that lack the natural permeability and water content found in conventional hydrothermal reservoirs. These formations are typically located at depths of 3-10 kilometers (1.9-6.2 miles) and can have temperatures exceeding 200°C (400°F).

- Reservoir Creation: To create an EGS reservoir, wells are drilled into the hot rock formation, and water is injected under high pressure to create a network of fractures. This process, known as hydraulic stimulation, enhances the permeability of the rock and allows the injected water to circulate and extract heat.
- Closed-Loop System: EGS operate as closed-loop systems, with the injected water continuously circulating through the engineered reservoir, extracting heat, and returning to the surface to drive turbines for electricity generation.
- Potential: EGS have the potential to significantly expand the geographic range of geothermal energy, as suitable hot rock formations are found in many regions worldwide. However, the technology is still in the development stage, with ongoing research aimed at optimizing reservoir creation and minimizing seismic risks.

Picture a remote, off-grid community located far from conventional geothermal resources. By developing an EGS project nearby, this community could access a reliable, baseload power source, reducing its dependence on diesel generators and improving its energy security.

3. Geothermal Heat Pumps
- Geothermal heat pumps, also known as ground-source heat pumps, are a versatile and efficient technology that uses the stable temperatures found just below the Earth's surface to provide heating, cooling, and hot water for buildings. Here are some key features of geothermal heat pumps:

- Resource Characteristics: Geothermal heat pumps rely on the relatively constant temperatures found a few meters below the Earth's surface. At these depths, the ground temperature remains between 10-16°C (50-60°F) throughout the year, providing a stable source of thermal energy.
- Heat Exchange: Geothermal heat pumps circulate a fluid (typically water or an antifreeze solution) through a network of underground pipes, known as a ground loop. In the winter, the fluid absorbs heat from the ground and carries it to the heat pump, which concentrates the heat and transfers it to the building. In the summer, the process is reversed, with the heat pump extracting heat from the building and transferring it to the ground.
- Efficiency: Geothermal heat pumps are highly efficient, as they move heat rather than generating it. They can provide 3-5 units of heating or cooling energy for every unit of electrical energy consumed, making them a cost-effective and environmentally friendly option for space conditioning.
- Wide Applicability: Geothermal heat pumps can be used in a wide range of settings, from residential homes to large commercial buildings. They can be installed in both new construction and retrofit projects, and their underground components have a long lifespan of 50 years or more.

Imagine a sustainable, net-zero energy home equipped with a geothermal heat pump. This home could maintain comfortable indoor temperatures year-round while minimizing its carbon footprint and energy costs, serving as a model for sustainable living.

As you can see, there are various types of geothermal systems, each with its own unique advantages and applications. Whether you're considering a large-scale geothermal power plant or a small-scale geothermal heat pump, understanding the different types of systems available can help you make an informed decision.

In the next section, we'll explore the advantages and disadvantages of geothermal energy, providing a balanced perspective on this reliable and sustainable renewable energy source.

Advantages and Disadvantages of Geothermal Energy

Geothermal energy offers a unique set of benefits and challenges compared to other renewable energy sources. In this section, we'll explore the advantages and disadvantages of geothermal energy to help you understand its potential role in the transition to a clean energy future.

Advantages of Geothermal Energy

1. Baseload Power: Unlike wind and solar energy, which are intermittent and depend on weather conditions, geothermal energy provides a reliable, baseload power source that can operate continuously, 24/7. This makes geothermal energy a valuable complement to other renewable energy sources and can help to balance the electricity grid.

2. Low Emissions: Geothermal power plants have very low greenhouse gas emissions compared to fossil fuel-based power plants. While some geothermal fluids may contain trace amounts of carbon dioxide or hydrogen sulfide, these emissions are generally much lower than those from coal or natural gas plants. Geothermal heat pumps, which use electricity to move heat, have an even lower environmental impact.

3. Small Land Footprint: Geothermal power plants have a relatively small land footprint compared to other energy sources. The underground components of geothermal systems, such as wells and pipelines, have minimal impact on the surface land use, and the aboveground facilities are compact. This makes geothermal energy a good option for areas with limited land availability.

4. Long Lifespan: Geothermal systems have a long operational lifespan. Geothermal power plants can operate for 30-50 years or more, while the underground components of geothermal heat pumps can last for 50 years or more. This long lifespan helps to offset the initial capital costs of geothermal projects and provides a stable, long-term energy supply.

5. Domestic Energy Source: Geothermal resources are widely distributed across the globe, and many countries have untapped geothermal potential. By developing these resources, countries can reduce their dependence on imported fossil fuels and improve their energy security.

6. Versatile Applications: Geothermal energy can be used for a wide range of applications beyond electricity generation, including space heating and cooling, hot water production, greenhouse heating, and industrial processes. This versatility makes geothermal energy a valuable resource for meeting diverse energy needs.

Imagine a city that has traditionally relied on coal-fired power plants for its electricity. By transitioning to geothermal energy, this city can enjoy cleaner air, a more stable energy supply, and new economic opportunities in the geothermal industry.

Disadvantages of Geothermal Energy

1. High Upfront Costs: Geothermal projects typically have high upfront capital costs, largely due to the expenses associated with exploration, drilling, and infrastructure development. These costs can be a barrier to entry for some developers and may require significant investment or financing.

2. Site-Specific Resource: Geothermal resources are site-specific and are not evenly distributed across the globe. The best geothermal resources are often found in areas with high heat flow, such as near tectonic plate boundaries or volcanic regions. This can limit the widespread adoption of geothermal energy in some areas.

3. Exploration Risk: Identifying and characterizing geothermal resources can be challenging and involves some exploration risk. Not all geothermal wells will be productive, and there is always the possibility that a geothermal project may not meet its expected capacity or performance.

4. Induced Seismicity: In some cases, geothermal development, particularly in EGS projects, can cause induced seismicity or small earthquakes. While these events are usually minor and can be managed through proper site selection and monitoring, they can raise concerns among local communities and regulators.

5. Water Use and Contamination: Geothermal power plants can require significant amounts of water for drilling, reservoir stimulation, and cooling. In areas with limited water resources, this can lead to competition with other water users. Additionally, if not properly managed, geothermal fluids can potentially contaminate groundwater resources.

6. Environmental Concerns: While geothermal energy is generally considered environmentally friendly, there are some potential impacts to consider. These may include the release of geothermal fluids containing trace amounts of greenhouse gases or other contaminants, as well as the visual impact of geothermal power plants in scenic areas.

Consider a scenario where a proposed geothermal project faces opposition due to concerns about induced seismicity and water use. In this case, developers would need to work closely with the local community and regulators to address these concerns, implement appropriate mitigation measures, and ensure the project's benefits outweigh its potential risks.

While geothermal energy offers significant advantages, it's essential to weigh them against the potential drawbacks and consider the specific circumstances of each project. By understanding both the pros and cons, policymakers, developers, and communities can make informed decisions about the role of geothermal energy in their energy mix.

As we move forward in exploring renewable energy solutions, it's crucial to consider the unique characteristics and potential of each technology, including geothermal energy. By taking a holistic and balanced approach, we can work towards a sustainable, low-carbon energy future that meets our diverse energy needs while minimizing our environmental impact.

Part II: Implementing Renewable Energy Solutions for Homes
Chapter 4:
Assessing Your Home's Energy Needs

Conducting an Energy Audit

Before implementing any renewable energy solutions in your home, it's essential to understand your household's energy needs and consumption patterns. By assessing your home's energy requirements, you can make informed decisions about which renewable energy technologies are best suited to your situation and how to optimize their performance. In this chapter, we'll explore the process of assessing your home's energy needs, starting with conducting an energy audit.

Conducting an Energy Audit
An energy audit is a comprehensive assessment of your home's energy usage, efficiency, and potential for improvement. By conducting an energy audit, you can identify areas where you can reduce energy waste, improve efficiency, and potentially integrate renewable energy solutions. Here's a step-by-step guide to conducting an energy audit in your home:

Step 1: Gather Your Energy Bills
Start by gathering your energy bills from the past year, including electricity, natural gas, and any other fuels you use in your home.

These bills will provide valuable information about your energy consumption patterns and costs.

Step 2: Identify High-Consumption Areas
Review your energy bills and identify the areas or appliances that consume the most energy in your home. Common high-consumption areas include heating and cooling systems, water heating, lighting, and large appliances like refrigerators and washing machines.

Step 3: Assess Your Home's Insulation and Air Sealing
Inspect your home's insulation and air sealing to determine if there are any gaps or inadequacies that could be leading to energy waste. Check the insulation in your walls, attic, and floors, and look for air leaks around windows, doors, and other openings.

Step 4: Evaluate Your Appliances and Electronics
Take an inventory of your appliances and electronics, noting their age, efficiency ratings, and usage patterns. Consider replacing old, inefficient appliances with newer, energy-efficient models to reduce your energy consumption.

Step 5: Examine Your Lighting
Assess your home's lighting fixtures and bulbs, and consider replacing inefficient incandescent bulbs with energy-saving LED or CFL bulbs. Also, evaluate your lighting usage patterns and identify areas where you can reduce unnecessary lighting or install motion sensors or timers.

Step 6: Consider Your Comfort and Habits
Reflect on your personal comfort preferences and energy usage habits. Are there areas where you can adjust your thermostat settings, reduce hot water usage, or unplug electronics when not in use to save energy without sacrificing comfort?

Step 7: Get a Professional Assessment
While a self-conducted energy audit can provide valuable insights, consider hiring a professional energy auditor for a more thorough and technical assessment. A professional auditor can use specialized tools and expertise to identify hidden energy waste and provide tailored recommendations for your home.

Imagine a family that has been struggling with high energy bills and discomfort in their home. By conducting an energy audit, they discover that their attic insulation is inadequate, leading to significant heat loss in the winter and heat gain in the summer. They also identify several appliances that are old and inefficient, as well as lighting fixtures that are wasting energy. Armed with this information, the family can make targeted improvements to their home's energy efficiency and explore renewable energy options that align with their needs and budget.

By conducting a comprehensive energy audit, you can gain a clear understanding of your home's energy needs, efficiency gaps, and potential for improvement. This information will be invaluable as you move forward in assessing your renewable energy potential and designing a tailored renewable energy solution for your home.

- Geothermal Potential: Consider the geological characteristics of your area, such as the presence of hot springs or other geothermal resources. Homes in areas with accessible geothermal resources have the potential to benefit from geothermal heating and cooling systems.

2. Evaluate Your Available Space and Resources
Consider the physical space and resources available on your property for implementing renewable energy technologies:

- Roof Space: Assess the size, orientation, and condition of your roof to determine its suitability for solar panel installation. A large, south-facing roof (in the Northern Hemisphere) with minimal shading is ideal for solar energy generation.

- Land Area: Evaluate the size and topography of your property to determine if you have sufficient space for ground-mounted solar panels, wind turbines, or geothermal systems. Large, open areas with minimal obstructions are best for these technologies.

- Access to Resources: Consider your access to the necessary resources for implementing renewable energy technologies, such as a suitable water source for geothermal systems or a stable soil structure for wind turbine foundations.

3. Consider Your Budget and Financial Incentives
Implementing renewable energy technologies in your home can involve significant upfront costs, so it's essential to consider your budget and available financial incentives:

- Upfront Costs: Research the upfront costs associated with the renewable energy technologies you are considering, including equipment, installation, and any necessary site preparations or upgrades.

- Long-Term Savings: Evaluate the potential long-term energy savings and financial returns of implementing renewable energy technologies in your home. Many renewable energy systems can provide significant savings on energy bills over their lifespan.

- Financial Incentives: Investigate the availability of financial incentives for renewable energy implementation in your area, such as government grants, tax credits, or net metering programs. These incentives can help offset the initial costs and improve the financial viability of your renewable energy project.

4. Consult with Renewable Energy Professionals

To get a more accurate assessment of your home's renewable energy potential, consider consulting with renewable energy professionals, such as:

- Renewable Energy Consultants: These experts can provide a comprehensive assessment of your home's renewable energy potential, recommend suitable technologies, and assist with project planning and implementation.

- Renewable Energy Installers: Experienced installers can provide practical insights into the feasibility and costs of implementing specific renewable energy technologies in your home, based on their local knowledge and expertise.

- Energy Auditors: If you haven't already conducted an energy audit, a professional energy auditor can help you identify energy efficiency opportunities and assess your home's suitability for renewable energy technologies.

Imagine a homeowner in a sunny, coastal region who has completed an energy audit and is now considering their renewable energy options. After assessing their location and climate, they determine that their home has excellent solar potential due to the abundant sunshine and minimal shading. They also have a large, south-facing roof that is well-suited for solar panel installation. The homeowner consults with a local solar installer and learns about the upfront costs, long-term savings, and available financial incentives for solar energy in their area. Based on this information, they decide to move forward with a rooftop solar panel system that will meet a significant portion of their home's energy needs and provide long-term financial benefits.

By thoroughly assessing your home's renewable energy potential, you can make informed decisions about which technologies are most suitable for your unique situation. This process will help you prioritize your renewable energy investments, maximize the benefits of your chosen technologies, and create a tailored renewable energy solution that meets your home's energy needs and aligns with your goals and budget.

Chapter 5
Designing a Solar Panel System for Your Home
Sizing Your Solar Panel Array

Once you have determined that your home has good potential for solar energy generation, the next step is to design a solar panel system that meets your energy needs and fits your budget. Designing a solar panel system involves several key considerations, including sizing your solar panel array, selecting the right components, and planning for installation and maintenance. In this chapter, we'll start by focusing on sizing your solar panel array.

Sizing Your Solar Panel Array
Sizing your solar panel array is one of the most critical aspects of designing a solar panel system for your home. A properly sized array will ensure that you generate enough electricity to meet your energy needs while optimizing your investment and available space. Here's a step-by-step guide to sizing your solar panel array:

Step 1: Determine Your Energy Needs
Start by reviewing your energy audit results and determining your average daily energy consumption in kilowatt-hours (kWh). You can calculate this by adding up your monthly energy usage from your utility bills and dividing by the number of days in the month.

For example, if your monthly energy usage is 900 kWh and there are 30 days in the month, your average daily energy consumption would be:
900 kWh ÷ 30 days = 30 kWh/day

Step 2: Assess Your Solar Potential
Next, assess your home's solar potential by considering factors such as your location, climate, and available roof or ground space. Determine the average daily peak sun hours in your area, which represents the number of hours per day that your solar panels will receive direct, unobstructed sunlight.

For instance, if you live in an area that receives an average of 5 peak sun hours per day, your solar panels will generate their rated output for approximately 5 hours each day.

Step 3: Calculate Your Solar Panel Array Size
To calculate the size of your solar panel array, divide your average daily energy consumption by your area's average daily peak sun hours. This will give you the total wattage of solar panels needed to meet your energy needs.

Using the example from Step 1 and assuming 5 peak sun hours per day, the calculation would be:
30 kWh/day ÷ 5 hours/day = 6 kW of solar panels needed

Step 4: Select Your Solar Panels
Choose solar panels that fit your energy needs, budget, and available space. Solar panels come in various wattages, typically ranging from 250 to 400 watts per panel. Divide the total wattage needed by the wattage of your chosen panels to determine the number of panels required.

For example, if you need a 6 kW array and choose 300-watt panels, you would need:
6,000 watts ÷ 300 watts/panel = 20 panels

Step 5: Consider System Losses and Overproduction
When sizing your solar panel array, it's essential to account for system losses due to factors such as temperature, shading, and inverter efficiency. To compensate for these losses, you may need to slightly oversize your array by 10-20%.

Additionally, consider your future energy needs and whether you want to overproduce electricity to offset energy consumption during cloudy days or to take advantage of net metering programs, which allow you to sell excess electricity back to the grid.

Step 6: Assess Your Available Space
Finally, assess your available roof or ground space to ensure that you have sufficient room for the number and size of panels needed. Consider factors such as roof orientation, tilt angle, and shading from nearby objects when determining the optimal placement for your panels.

If you have limited space, you may need to adjust your array size or consider higher-efficiency panels that generate more electricity per square foot.

Imagine a homeowner who has determined that their average daily energy consumption is 25 kWh and they live in an area with 5.5 peak sun hours per day. They choose 350-watt solar panels and account for a 15% system loss.

The calculation for their solar panel array size would be:
(25 kWh/day ÷ 5.5 hours/day) × 1.15 = 5.23 kW of solar panels needed
5,230 watts ÷ 350 watts/panel ≈ 15 panels needed

The homeowner then assesses their south-facing roof and determines that they have sufficient space for 15 panels, taking into account the optimal tilt angle and spacing between panels.

By properly sizing your solar panel array, you can ensure that your solar panel system generates sufficient electricity to meet your energy needs while optimizing your investment and available space. Keep in mind that sizing your array is just one aspect of designing a complete solar panel system.

In the next section, we'll explore how to choose the right components for your solar panel system, including inverters, batteries, and mounting hardware, to ensure optimal performance and reliability.

Choosing the Right Components

After sizing your solar panel array, the next step in designing your solar panel system is selecting the right components to ensure optimal performance, reliability, and safety. The main components of a solar panel system include solar panels, inverters, mounting hardware, and, optionally, batteries for energy storage. In this section, we'll discuss how to choose the right components for your home solar panel system.

1. Solar Panels

When selecting solar panels for your system, consider the following factors:

- Efficiency: Choose solar panels with high efficiency ratings to maximize energy generation per square foot. Higher efficiency panels may cost more upfront but can generate more electricity in a smaller space.

- Durability: Look for solar panels with robust construction and long warranties to ensure they can withstand harsh weather conditions and maintain performance over their lifespan.

- Type: Decide between monocrystalline, polycrystalline, or thin-film solar panels based on your efficiency needs, space constraints, and budget.

2. Inverters

Inverters convert the direct current (DC) electricity generated by your solar panels into alternating current (AC) electricity that can be used in your home. When choosing an inverter, consider:

- Type: There are two main types of inverters: string inverters and microinverters. String inverters are centralized and connect multiple solar panels in series, while microinverters are mounted on each individual panel. Microinverters offer better performance in shaded conditions and allow for panel-level monitoring, but they may be more expensive.

- Efficiency: Look for inverters with high efficiency ratings to minimize energy losses during the DC to AC conversion process.

- Compatibility: Ensure that your chosen inverter is compatible with your solar panels and can handle the total wattage of your array.

3. Mounting Hardware

Mounting hardware secures your solar panels to your roof or ground-mounted structure. When selecting mounting hardware, consider:

- Compatibility: Choose mounting hardware that is compatible with your solar panels and can accommodate their size and weight.

- Durability: Look for mounting hardware made from corrosion-resistant materials, such as aluminum or stainless steel, to withstand harsh weather conditions and ensure long-term stability.

- Adjustability: Consider mounting hardware that allows for adjustable tilt angles to optimize solar panel orientation and maximize energy generation.

4. Batteries (Optional)

If you want to store excess solar energy for use during low-production periods or grid outages, you'll need to include batteries in your solar panel system. When choosing batteries, consider:

- Type: There are several types of batteries suitable for solar energy storage, including lead-acid, lithium-ion, and saltwater batteries. Each type has its own advantages and disadvantages in terms of cost, lifespan, and maintenance requirements.

- Capacity: Choose a battery capacity that meets your energy storage needs and budget. Consider factors such as your daily energy consumption, desired backup power duration, and available space for battery storage.

- Compatibility: Ensure that your chosen batteries are compatible with your inverter and can be safely integrated into your solar panel system.

5. Monitoring Systems

Monitoring systems allow you to track your solar panel system's performance, energy generation, and consumption. When selecting a monitoring system, look for:

- Real-time data: Choose a system that provides real-time data on your solar panel system's performance, allowing you to identify any issues or inefficiencies quickly.

- Remote access: Consider a monitoring system with remote access capabilities, enabling you to monitor your system from anywhere using a smartphone or web-based platform.

- Compatibility: Ensure that your monitoring system is compatible with your inverter and other system components.

Imagine a homeowner who has chosen high-efficiency monocrystalline solar panels for their 6 kW array. They select a string inverter with a 6 kW capacity and 98% efficiency rating, along with aluminum mounting hardware that allows for adjustable tilt angles. To provide backup power during grid outages, they also include a 10 kWh lithium-ion battery system compatible with their inverter. Finally, they install a monitoring system that provides real-time performance data and remote access via a smartphone app.

By carefully selecting the right components for your solar panel system, you can ensure optimal performance, reliability, and safety while meeting your specific energy needs and budget. Keep in mind that the quality and compatibility of your system components are just as important as proper sizing and installation.

Installation and Maintenance

Once you have sized your solar panel array and chosen the right components, the next step is to install and maintain your home solar panel system. Proper installation and regular maintenance are essential for ensuring the safety, performance, and longevity of your system. In this section, we'll discuss the key considerations for installation and maintenance of your home solar panel system.

Installation

1. Professional Installation

It is strongly recommended to have your solar panel system installed by a licensed and experienced professional. A professional installer will ensure that your system is designed and installed according to local building codes, electrical standards, and industry best practices. They will also handle necessary permits and inspections, and provide warranties for their work.

2. Site Preparation

Before installation, your installer will prepare your site by assessing your roof or ground-mounted structure, determining the optimal panel layout, and making any necessary structural reinforcements or upgrades. They will also install the mounting hardware and any required electrical components, such as wiring, conduit, and disconnects.

3. Panel and Inverter Installation

Your installer will then mount and secure your solar panels to the mounting hardware, ensuring proper alignment and spacing for optimal performance. They will connect the panels to your inverter(s) using appropriate wiring and connectors, following manufacturer guidelines and electrical codes.

4. Battery and Monitoring System Installation

If your system includes batteries for energy storage, your installer will safely connect them to your inverter and ensure proper charging and discharging functions. They will also install your monitoring system, configuring it to track your system's performance and energy generation.

5. System Testing and Commissioning

After installation, your installer will thoroughly test your solar panel system to ensure all components are functioning properly and safely. They will also commission your system, connecting it to the electrical grid (if applicable) and obtaining final approvals from your local utility and building department.

Maintenance

1. Regular Inspections

To ensure your solar panel system continues to perform optimally and safely, it is important to conduct regular inspections. Inspect your panels, mounting hardware, and electrical components for any signs of damage, corrosion, or wear. Check for any shading or debris on your panels that may reduce energy generation.

2. Cleaning

Over time, dust, dirt, and other debris can accumulate on your solar panels, reducing their efficiency. Regularly clean your panels using a soft brush or cloth and mild detergent, following manufacturer guidelines. Avoid using abrasive materials or high-pressure water, which can damage your panels.

3. Inverter and Battery Maintenance

Monitor your inverter and battery performance using your monitoring system, and schedule any necessary maintenance or repairs. Inverters typically have a lifespan of 10-15 years and may require replacement. Batteries also have a limited lifespan and may need to be replaced every 5-10 years, depending on the type and usage.

4. Professional Service

Schedule professional maintenance services every 1-2 years to ensure your system is operating safely and efficiently. A professional technician can conduct thorough inspections, identify any potential issues, and make necessary repairs or adjustments.

5. Monitoring and Optimization

Regularly monitor your solar panel system's performance using your monitoring system, and optimize your energy usage and storage to maximize the benefits of your system. Track your energy generation and consumption patterns, and adjust your usage habits or battery storage settings to minimize reliance on the grid during peak periods.

Imagine a homeowner who has hired a professional installer to design and install their 6 kW solar panel system. The installer prepares the site, mounts the panels and inverter, and connects the battery and monitoring system. After thorough testing and commissioning, the system is approved and begins generating clean energy for the home. The homeowner conducts regular inspections and cleaning, and schedules annual professional maintenance to ensure optimal performance and longevity of their system.

By prioritizing proper installation and maintenance of your home solar panel system, you can maximize its safety, performance, and lifespan, ensuring a reliable and sustainable source of clean energy for your home. Remember that investing in quality installation and regular maintenance can save you costly repairs and energy losses in the long run.

As you embark on your journey to harness solar energy for your home, keep in mind the importance of proper sizing, component selection, installation, and maintenance. By following these guidelines and working with experienced professionals, you can create a custom solar panel system that meets your energy needs, budget, and sustainability goals, while providing long-term benefits for your home and the environment.

Chapter 6
Designing a Wind Turbine System for Your Home

Assessing Wind Resources in Your Area

After exploring solar energy options for your home, let's now consider another renewable energy source: wind power. Designing a wind turbine system for your home involves several key steps, including assessing your wind resources, selecting the right turbine size and type, and planning for installation and maintenance. In this chapter, we'll start by focusing on assessing the wind resources in your area.

Assessing Wind Resources in Your Area
Before investing in a wind turbine system for your home, it's crucial to determine whether your location has sufficient wind resources to generate a viable amount of energy. Assessing your wind resources involves evaluating wind speed, direction, and consistency, as well as considering any potential obstacles or restrictions. Here's a step-by-step guide to assessing wind resources in your area:

Step 1: Determine Average Wind Speed
The first step in assessing your wind resources is to determine the average wind speed in your location. Wind speed is the most critical factor in determining the feasibility and productivity of a wind turbine system. Generally, an average wind speed of at least 5 meters per second (11 miles per hour) is required for a small wind turbine to generate a significant amount of energy.

To determine your average wind speed, you can:
- Consult wind resource maps: Many countries have wind resource maps that provide estimated average wind speeds for different regions. These maps can give you a general idea of the wind potential in your area.
- Use online tools: Some online tools, such as the Global Wind Atlas or the National Renewable Energy Laboratory's (NREL) Wind Resource Assessment Tool, allow you to input your location and get an estimate of your average wind speed.
- Conduct on-site measurements: For the most accurate assessment, you can install an anemometer (wind speed meter) at your proposed turbine location and collect data over several months to a year. This will give you a clear picture of your specific wind resource potential.

Step 2: Assess Wind Direction and Consistency

In addition to wind speed, it's important to consider wind direction and consistency when assessing your wind resources. Ideally, you want your wind turbine to be exposed to consistent winds from a predominant direction, as this will maximize energy generation and minimize turbulence and wear on the turbine.

To assess wind direction and consistency, you can:
- Observe local wind patterns: Pay attention to local wind patterns, such as seasonal variations and daily wind cycles. Note any predominant wind directions and how consistent the winds are throughout the year.

- Use a wind vane: A wind vane can help you determine the prevailing wind direction at your location. Record the wind direction data along with your wind speed measurements to get a complete picture of your wind resources.
- Consult local weather data: Local weather stations or airports may have historical wind data for your area, including average wind speeds and prevailing wind directions.

Step 3: Consider Obstacles and Restrictions
When assessing your wind resources, it's important to consider any potential obstacles or restrictions that may impact your wind turbine's performance or installation. These may include:

- Zoning regulations: Check with your local zoning authority to ensure that wind turbines are permitted in your area and to understand any height, setback, or noise restrictions.
- Proximity to neighbors: Consider the proximity of your proposed turbine location to your neighbors' properties, as well as any potential visual or noise impacts.
- Obstacles: Assess any obstacles, such as trees, buildings, or hills, that may create turbulence or reduce wind speed at your proposed turbine location. Ideally, your turbine should be sited at least 10 meters (33 feet) above any obstacles within 100 meters (330 feet).

Step 4: Estimate Energy Production

Once you have assessed your wind speed, direction, and consistency, and considered any potential obstacles or restrictions, you can estimate the potential energy production of a wind turbine at your location. Many wind turbine manufacturers provide energy production estimates based on average wind speeds, which can help you determine the feasibility and cost-effectiveness of a wind turbine system for your home.

Imagine a homeowner living in a rural area with an average wind speed of 6 meters per second (13.4 miles per hour), consistent winds from the southwest, and no significant obstacles or restrictions. They install an anemometer at their proposed turbine location and collect data for six months, confirming their initial assessment. Based on this information, they estimate that a small wind turbine could generate a significant portion of their home's energy needs, making it a viable renewable energy option.

By thoroughly assessing your wind resources, you can make an informed decision about whether a wind turbine system is a practical and cost-effective solution for your home. Keep in mind that wind resource assessment is just the first step in designing a wind turbine system, and there are many other factors to consider, such as turbine size, type, and installation requirements.

In the next section, we'll discuss how to choose the right wind turbine for your home based on your energy needs, wind resources, and budget.

Choosing the Right Wind Turbine

After assessing your wind resources and determining that your location is suitable for a wind turbine system, the next step is to choose the right wind turbine for your home. Selecting the appropriate wind turbine involves considering factors such as energy needs, turbine size, type, and efficiency, as well as your budget and maintenance requirements. In this section, we'll discuss how to choose the right wind turbine for your home.

1. Determine Your Energy Needs

The first step in choosing the right wind turbine is to determine your home's energy needs. Review your energy audit results and calculate your average daily energy consumption in kilowatt-hours (kWh). This will help you determine the size and capacity of the wind turbine needed to meet your energy requirements.

2. Consider Turbine Size and Capacity

Wind turbines come in various sizes and capacities, typically ranging from small, residential-scale turbines (1-10 kW) to larger, community-scale turbines (10-100 kW). When selecting a wind turbine size, consider the following factors:

- Energy production: Choose a turbine size that can generate enough energy to meet your needs based on your wind resource assessment and energy consumption.
- Available space: Consider the space available for your wind turbine installation, including any zoning restrictions or setback requirements.

- Tower height: Generally, taller towers provide access to stronger and more consistent winds. Choose a turbine size that is appropriate for the tower height you can accommodate on your property.

3. Evaluate Turbine Types

There are two main types of wind turbines: horizontal-axis wind turbines (HAWTs) and vertical-axis wind turbines (VAWTs). HAWTs are the most common type and are typically more efficient, while VAWTs are less common but can be suitable for areas with turbulent or variable winds.

When choosing a turbine type, consider the following factors:
- Efficiency: HAWTs are generally more efficient than VAWTs, as they can capture more energy from the wind due to their design and orientation.
- Wind conditions: VAWTs may be more suitable for areas with turbulent or variable winds, as they can capture wind from any direction without the need for a yaw mechanism to orient the turbine into the wind.
- Maintenance: HAWTs typically require more maintenance than VAWTs due to their complex gearboxes and yaw mechanisms. However, VAWTs may have a shorter lifespan due to the higher stress on their bearings and components.

4. Compare Turbine Efficiency and Performance
- When choosing a wind turbine, compare the efficiency and performance of different models to ensure you select the best option for your needs. Key factors to consider include:

- Power coefficient: The power coefficient (Cp) is a measure of how efficiently a turbine converts wind energy into electrical energy. Higher Cp values indicate more efficient turbines.
- Rated power: The rated power is the maximum power output of the turbine at a specified wind speed. Choose a turbine with a rated power that aligns with your energy needs and wind resources.
- Cut-in and cut-out wind speeds: The cut-in speed is the minimum wind speed at which the turbine starts generating power, while the cut-out speed is the maximum wind speed at which the turbine shuts down to prevent damage. Ensure that your chosen turbine's cut-in and cut-out speeds are appropriate for your location's wind conditions.

5. Consider Budget and Maintenance

Finally, consider your budget and the maintenance requirements of your chosen wind turbine. While larger, more efficient turbines may have a higher upfront cost, they can generate more energy and provide a better return on investment over time. Additionally, factor in the cost of regular maintenance, such as annual inspections, blade cleaning, and bearing replacements, to ensure your turbine operates safely and efficiently throughout its lifespan.

Imagine a homeowner who has determined that their daily energy consumption is 30 kWh and their location has an average wind speed of 6 meters per second. They have enough space to accommodate a small, residential-scale HAWT with a rated power of 5 kW.

After comparing the efficiency and performance of several models, they choose a turbine with a high Cp value, a cut-in speed of 3 meters per second, and a cut-out speed of 25 meters per second, which aligns well with their location's wind conditions. The homeowner also factors in the cost of regular maintenance and plans for annual inspections and repairs to keep their turbine running smoothly.

By carefully choosing the right wind turbine for your home based on your energy needs, wind resources, and budget, you can maximize the benefits of your wind energy system and ensure a reliable, sustainable source of clean energy. Keep in mind that proper sizing, efficiency, and maintenance are key to the long-term success and cost-effectiveness of your wind turbine installation.

In the next section, we'll discuss the installation and maintenance requirements for your home wind turbine system, including site preparation, tower selection, and regular upkeep to ensure optimal performance and safety.

INSTALLATION AND MAINTENANCE

Once you have chosen the right wind turbine for your home, the next step is to plan for its installation and maintain it properly to ensure optimal performance, safety, and longevity. Installing a wind turbine involves several key considerations, such as site preparation, tower selection, and electrical connections, while regular maintenance is essential to keep your turbine running efficiently and safely. In this section, we'll discuss the installation and maintenance requirements for your home wind turbine system.

Installation

1. Site Preparation
Before installing your wind turbine, you'll need to prepare the site to ensure a stable, secure, and efficient installation. This includes:

- Clearing the area: Remove any obstacles or vegetation that may interfere with the turbine's operation or access for maintenance.
- Grading and leveling: Ensure that the site is graded and leveled to provide a stable foundation for the turbine tower.
- Concrete foundation: Pour a concrete foundation according to the manufacturer's specifications and local building codes to anchor the turbine tower securely.

2. Tower Selection and Installation

The tower is a critical component of your wind turbine system, as it provides the necessary height to access strong, consistent winds. When selecting and installing your tower, consider the following factors:

- Tower height: Choose a tower height that optimizes wind speed and energy production while complying with local zoning restrictions and safety requirements.
- Tower type: There are two main types of towers: guyed towers and self-supporting towers. Guyed towers are anchored by cables and are typically less expensive, while self-supporting towers are freestanding and require a larger foundation.
- Material: Towers can be made from various materials, such as steel or aluminum. Choose a material that is durable, corrosion-resistant, and suitable for your local climate and wind conditions.
- Installation: Follow the manufacturer's instructions and local building codes when erecting the tower, ensuring that it is plumb, secure, and properly anchored to the foundation.

3. Turbine and Electrical Installation

Once the tower is installed, you can mount the wind turbine and connect it to your home's electrical system. This process includes:

- Turbine assembly: Assemble the wind turbine according to the manufacturer's instructions, ensuring that all components are properly aligned and secured.

- Turbine mounting: Lift the assembled turbine to the top of the tower using a crane or other suitable equipment, and secure it to the tower according to the manufacturer's specifications.
- Electrical connections: Connect the turbine to your home's electrical system through a suitable inverter and battery bank (if applicable), following the manufacturer's wiring diagrams and local electrical codes.
- Grounding and lightning protection: Ensure that the turbine and tower are properly grounded and equipped with lightning protection to minimize the risk of damage from electrical surges or strikes.

4. Commissioning and Testing

After installation, commission and test your wind turbine system to ensure that it is operating safely and efficiently. This includes:

- Visual inspection: Conduct a thorough visual inspection of the turbine, tower, and electrical connections to identify any defects, loose components, or potential hazards.
- Performance testing: Test the turbine's power output, voltage, and current to ensure that it is generating energy as expected and within the manufacturer's specifications.
- Safety checks: Verify that all safety features, such as overspeed protection, yaw control, and emergency braking, are functioning properly.

Maintenance

1. Regular Inspections

To keep your wind turbine system operating safely and efficiently, conduct regular inspections at least once a year, or more frequently if recommended by the manufacturer. During these inspections:

- Check for any visible damage, corrosion, or wear on the turbine blades, hub, and tower.
- Inspect electrical connections, wiring, and grounding for any signs of damage, loose connections, or corrosion.
- Monitor the turbine's performance using the manufacturer's software or monitoring system to identify any deviations from normal operation.

2. Lubrication and Cleaning

Regularly lubricate and clean your wind turbine components to prevent premature wear and maintain optimal performance. This includes:

- Greasing bearings and gears according to the manufacturer's recommendations to reduce friction and wear.
- Cleaning the turbine blades to remove any dirt, debris, or ice accumulation that may affect the turbine's efficiency and balance.

3. Component Replacement and Repairs

Over time, certain wind turbine components may require replacement or repair due to normal wear and tear. Common components that may need attention include:

- Blades: Replace or repair any damaged, cracked, or imbalanced blades to maintain the turbine's efficiency and prevent further damage.
- Bearings and gears: Replace worn bearings and gears to ensure smooth, efficient operation and prevent premature failure.
- Electrical components: Replace or repair any faulty electrical components, such as inverters, controllers, or wiring, to maintain the system's safety and reliability.

4. Professional Maintenance and Support

While some maintenance tasks can be performed by the homeowner, it's essential to have a professional wind turbine technician conduct regular maintenance and repairs, especially for more complex or high-risk tasks. Establish a relationship with a qualified service provider who can offer ongoing support, troubleshooting, and maintenance for your wind turbine system.

Imagine a homeowner who has installed a 5 kW HAWT on a 30-meter guyed tower on their property. They have followed all the necessary site preparation, installation, and commissioning steps, and the turbine is now generating clean energy for their home. To ensure the system's longevity and performance, they conduct monthly visual inspections, clean the blades annually, and lubricate the bearings every six months. They also have a professional technician perform a comprehensive annual maintenance check and address any issues that arise, such as replacing a worn bearing or updating the turbine's control software.

By prioritizing proper installation and maintenance of your home wind turbine system, you can maximize its safety, efficiency, and lifespan, ensuring a reliable and sustainable source of clean energy for your home. Remember that investing in quality components, professional installation, and regular maintenance can save you costly repairs and energy losses in the long run.

As you embark on your journey to harness wind energy for your home, keep in mind the importance of site assessment, turbine selection, installation, and maintenance. By following these guidelines and working with experienced professionals, you can create a custom wind turbine system that meets your energy needs, budget, and sustainability goals, while providing long-term benefits for your home and the environment.

Chapter 7
Designing a Geothermal System for Your Home
Evaluating Your Home's Geothermal Potential

Geothermal energy is a clean, reliable, and efficient renewable energy source that can provide significant benefits for businesses looking to reduce their carbon footprint and energy costs. Designing a geothermal system for your business involves several key steps, including evaluating your geothermal potential, choosing the right system type and size, and planning for installation and maintenance. In this chapter, we'll start by focusing on evaluating your business's geothermal potential.

Evaluating Your Business's Geothermal Potential
Before investing in a geothermal system for your business, it's essential to assess whether your location has suitable geothermal resources and whether a geothermal system is a feasible and cost-effective solution for your energy needs. Evaluating your business's geothermal potential involves considering factors such as geology, land availability, and energy requirements. Here's a step-by-step guide to evaluating your business's geothermal potential:

Step 1: Assess Geological Conditions

The first step in evaluating your geothermal potential is to assess the geological conditions at your location. Geothermal resources are most accessible in areas with high heat flow, such as near tectonic plate boundaries, volcanic regions, or deep sedimentary basins. Factors to consider include:

- Geothermal gradient: The geothermal gradient is the rate at which temperature increases with depth in the Earth's crust. Areas with higher geothermal gradients are more suitable for geothermal energy development.
- Rock type and permeability: The type and permeability of the rock formations at your location can impact the feasibility and cost of drilling geothermal wells and extracting heat.
- Groundwater availability: The presence of groundwater can enhance heat transfer and improve the efficiency of geothermal systems. Assess the depth, temperature, and flow rate of any aquifers at your location.

To gather this information, consult local geological maps, geothermal resource assessments, and geothermal energy experts who can provide site-specific evaluations.

Step 2: Evaluate Land Availability and Accessibility

Next, evaluate the land availability and accessibility at your business location. Geothermal systems require sufficient space for drilling wells, installing heat exchangers, and constructing any necessary surface facilities. Consider the following factors:

- Land ownership: Ensure that you have the necessary rights and permissions to develop geothermal resources on your property or any adjacent land.
- Drilling access: Assess the accessibility of your site for drilling rigs and other equipment needed to construct geothermal wells. Consider any logistical challenges, such as steep terrain or limited road access.
- Surface facilities: Determine the space requirements for any surface facilities, such as heat exchangers, pumps, and piping, and ensure that you have adequate room for their installation and maintenance.

Step 3: Determine Energy Requirements and Efficiency

To properly size your geothermal system, you'll need to determine your business's energy requirements and the efficiency of the proposed system. This involves:

- Energy audit: Conduct a comprehensive energy audit to assess your current energy consumption, including heating, cooling, and electricity needs. This will help you determine the required capacity of your geothermal system.
- System efficiency: Evaluate the efficiency of different geothermal system types and configurations to determine which option best meets your energy needs and maximizes the use of your geothermal resources. Factors to consider include heat exchanger design, pumping requirements, and distribution systems.
- Cost-benefit analysis: Perform a cost-benefit analysis to compare the upfront costs and long-term savings of a geothermal system against conventional energy sources.

- Consider factors such as installation costs, operating and maintenance expenses, and potential energy savings over the system's lifespan.

Step 4: Consult with Geothermal Professionals
To ensure a thorough and accurate evaluation of your geothermal potential, consult with experienced geothermal professionals, such as geologists, engineers, and system designers. These experts can provide valuable insights and recommendations based on your specific location, energy needs, and business goals. They can also assist with site assessments, system design, and cost estimations.

Imagine a manufacturing facility located near a known geothermal resource area. The facility conducts a thorough geological assessment and determines that the site has a high geothermal gradient and suitable rock formations for drilling. They also evaluate their land availability and find sufficient space for the necessary wells and surface facilities. After conducting an energy audit and cost-benefit analysis, they determine that a geothermal system can provide a significant portion of their heating and cooling needs while reducing their long-term energy costs. The facility then engages a team of geothermal professionals to develop a detailed system design and implementation plan.

By carefully evaluating your business's geothermal potential, you can make an informed decision about whether a geothermal system is a practical and cost-effective solution for your energy needs. Keep in mind that a thorough assessment is crucial for the success and long-term viability of your geothermal project.

Choosing the Right Geothermal System

Once you have evaluated your business's geothermal potential and determined that a geothermal system is a viable option, the next step is to choose the right system for your commercial needs. Selecting the appropriate geothermal system involves considering factors such as system type, size, efficiency, and compatibility with your existing infrastructure. In this section, we'll discuss how to choose the right commercial geothermal system for your business.

1. Determine System Type
There are three main types of commercial geothermal systems: direct use, ground source heat pumps (GSHPs), and enhanced geothermal systems (EGS). Consider the following factors when determining the most suitable system type for your business:

- Direct use systems: These systems use geothermal resources directly for heating or cooling applications, such as space heating, water heating, or industrial processes. Direct use systems are most suitable for businesses with access to high-temperature geothermal resources and a consistent demand for heat.
- Ground source heat pumps (GSHPs): GSHPs use the relatively constant temperature of the Earth's shallow subsurface to provide heating and cooling. They are the most common type of commercial geothermal system and are suitable for a wide range of businesses, particularly those with balanced heating and cooling needs.

- Enhanced geothermal systems (EGS): EGS involve creating an artificial geothermal reservoir by fracturing hot rock formations and circulating water through the created fractures. These systems are most suitable for businesses with access to deep, high-temperature geothermal resources but limited natural permeability.

2. Evaluate System Size and Capacity

The size and capacity of your geothermal system will depend on your business's energy requirements, as determined by your energy audit. When evaluating system size and capacity, consider the following factors:

- Peak load: Determine the peak heating and cooling load of your business to ensure that the geothermal system can meet your maximum energy demand.
- Annual energy consumption: Calculate your annual energy consumption for heating and cooling to properly size your geothermal system for optimal performance and efficiency.
- Load profile: Analyze your business's daily and seasonal load profile to identify any variations in energy demand that may impact system sizing and design.
- Redundancy and backup: Consider incorporating redundancy or backup systems to ensure uninterrupted operation in case of maintenance or equipment failure.

3. Assess System Efficiency and Performance

When choosing a commercial geothermal system, assess the efficiency and performance of different options to maximize energy savings and minimize operating costs. Key factors to consider include:

- Coefficient of Performance (COP): The COP is a measure of a geothermal system's efficiency in providing heating or cooling. Higher COPs indicate more efficient systems that can deliver more energy output per unit of input.
- Energy Efficiency Ratio (EER): The EER is a measure of a geothermal system's cooling efficiency, calculated as the ratio of cooling capacity to power input. Higher EERs indicate more efficient cooling performance.
- Heat exchanger effectiveness: The effectiveness of the heat exchanger in transferring heat between the geothermal fluid and the building's heating and cooling system can significantly impact overall system efficiency.
- Pumping requirements: Evaluate the pumping requirements for circulating geothermal fluid and consider the associated energy consumption and operating costs.

4. Consider Integration and Compatibility

When selecting a commercial geothermal system, consider its integration and compatibility with your existing building infrastructure and energy systems. Factors to evaluate include:

- HVAC system compatibility: Ensure that the geothermal system is compatible with your existing heating, ventilation, and air conditioning (HVAC) equipment, such as air handlers, ductwork, and control systems.
- Electrical system compatibility: Verify that your building's electrical system can accommodate the power requirements of the geothermal system, including any necessary upgrades or modifications.

- Spatial requirements: Consider the spatial requirements for the geothermal system components, such as heat exchangers, pumps, and piping, and ensure that your facility has adequate space for their installation and maintenance.

5. Evaluate Costs and Return on Investment

Finally, evaluate the costs and return on investment (ROI) of different commercial geothermal system options to determine the most cost-effective solution for your business. Consider the following factors:

- Initial installation costs: Obtain detailed cost estimates for the installation of the geothermal system, including well drilling, heat exchangers, pumps, piping, and any necessary building modifications.
- Operating and maintenance costs: Estimate the ongoing operating and maintenance costs of the geothermal system, including electricity consumption, regular maintenance, and any potential repairs or replacements.
- Energy savings and ROI: Calculate the projected energy savings and ROI of the geothermal system based on your current energy costs, system efficiency, and estimated lifespan. Compare these savings to the upfront installation and ongoing operating costs to determine the financial viability of the project.

Imagine a large office complex that has determined that a ground source heat pump system is the most suitable geothermal option for their energy needs. They work with a geothermal system designer to properly size the system based on their peak load, annual energy consumption, and load profile.

The designer recommends a high-efficiency heat pump with a COP of 4.5 and an EER of 20, coupled with a closed-loop ground heat exchanger. The system is designed to integrate seamlessly with the building's existing HVAC infrastructure and electrical system. After evaluating the installation and operating costs, the office complex determines that the geothermal system will provide a significant return on investment through reduced energy costs and lower maintenance requirements over its 25-year lifespan.

By carefully choosing the right commercial geothermal system based on your business's specific energy requirements, geothermal resources, and budget, you can maximize the benefits of this clean, reliable, and efficient renewable energy source. Keep in mind that proper system selection, sizing, and design are critical for the long-term performance, cost-effectiveness, and sustainability of your geothermal investment.

In the next section, we'll discuss the installation, maintenance, and monitoring considerations for commercial geothermal systems to ensure optimal performance and longevity.

Installation and Maintenance

After selecting the right commercial geothermal system for your business, the next crucial steps are proper installation, regular maintenance, and ongoing monitoring to ensure optimal performance, efficiency, and longevity. Installing a geothermal system involves several key considerations, such as site preparation, drilling, and system integration, while regular maintenance and monitoring are essential for identifying and addressing any issues that may arise. In this section, we'll discuss the installation, maintenance, and monitoring requirements for your commercial geothermal system.

Installation

1. Site Preparation and Permitting

Before installing your geothermal system, properly prepare the site and obtain all necessary permits and approvals. This includes:

- Conducting a detailed site survey to identify any potential obstacles or constraints, such as underground utilities or environmental sensitivities.
- Obtaining all required permits and approvals from local, state, and federal authorities, such as building permits, drilling permits, and environmental assessments.
- Preparing the site for drilling and installation, including clearing and grading the area, establishing access roads, and setting up any necessary safety measures.

2. Drilling and Well Installation

The drilling and installation of geothermal wells is a critical component of the installation process. This involves:

- Drilling the geothermal wells to the required depth and diameter based on the system design and geological conditions. This may involve using specialized drilling rigs and techniques, such as directional drilling or casing installation.
- Installing the well casing, grouting, and any necessary well screens or packers to ensure the integrity and longevity of the well.
- Conducting well logging and testing to verify the well's integrity, productivity, and thermal properties.

3. Heat Exchanger and Piping Installation

After the wells are installed, the heat exchanger and piping system must be installed to transfer heat between the geothermal fluid and the building's HVAC system. This includes:

- Installing the heat exchanger, circulation pumps, and any necessary valves or controls based on the system design and manufacturer's specifications.
- Connecting the heat exchanger to the geothermal well piping and the building's HVAC system, ensuring proper insulation and sealing to minimize heat loss.
- Pressure testing and flushing the piping system to verify its integrity and remove any debris or contaminants.

4. System Integration and Commissioning

The final step in the installation process is integrating the geothermal system with the building's existing HVAC and electrical systems and commissioning the system for operation. This involves:

- Connecting the geothermal system to the building's HVAC system, including any necessary modifications or upgrades to the existing ductwork, piping, or control systems.
- Installing and programming the system controls, monitoring equipment, and any necessary safety devices or alarms.
- Conducting a thorough commissioning process to verify that the system is operating as designed, including testing and balancing the system, calibrating sensors and controls, and providing operator training.

Maintenance and Monitoring

1. Regular Inspection and Testing

To ensure the ongoing performance and efficiency of your commercial geothermal system, conduct regular inspections and testing, including:

- Visual inspections of the wells, piping, heat exchanger, and other system components to identify any signs of wear, damage, or leakage.
- Pressure and flow testing to verify that the system is operating within the designed parameters and to detect any blockages or restrictions.

- Water quality testing to monitor the geothermal fluid's chemistry and to identify any potential scaling, corrosion, or biological growth issues.

2. Preventive Maintenance

Implement a comprehensive preventive maintenance program to proactively address any potential issues and extend the lifespan of your geothermal system. This may include:

- Regularly cleaning and flushing the heat exchanger and piping system to remove any accumulated debris or scale.
- Lubricating and replacing any worn or damaged components, such as pumps, valves, or bearings, based on the manufacturer's recommendations.
- Updating and calibrating the system controls and monitoring equipment to ensure optimal performance and efficiency.

3. Performance Monitoring and Optimization

Continuously monitor and optimize your geothermal system's performance to maximize energy savings and minimize operating costs. This involves:

- Installing and utilizing a comprehensive monitoring system to track key performance indicators, such as energy consumption, system efficiency, and operating temperatures.
- Analyzing performance data to identify any trends, anomalies, or opportunities for improvement, such as adjusting system setpoints or reconfiguring piping layouts.

- Implementing any necessary system modifications or upgrades to optimize performance and efficiency based on the monitoring data and analysis.

4. Emergency Response and Troubleshooting

Establish a clear protocol for responding to any system emergencies or malfunctions and troubleshooting any issues that arise. This includes:

- Developing a detailed emergency response plan that outlines the procedures for shutting down the system, isolating any damaged components, and notifying relevant personnel.
- Training facility staff and maintenance personnel on the proper operation and troubleshooting of the geothermal system, including identifying and diagnosing common issues.
- Establishing a network of qualified geothermal service providers and equipment suppliers to quickly respond to any emergencies or repair needs.

Imagine a hospital that has installed a large-scale ground source heat pump system to provide heating and cooling for its facilities. The installation process involved careful site preparation, including a detailed site survey and obtaining all necessary permits and approvals. The geothermal wells were drilled to a depth of 500 feet using specialized directional drilling techniques, and the well casings were installed and grouted to ensure long-term integrity. The heat exchanger and piping system were installed and integrated with the hospital's existing HVAC system, and the entire system was thoroughly commissioned to verify optimal performance.

The hospital implements a comprehensive maintenance and monitoring program, including regular inspections, preventive maintenance, and continuous performance monitoring, to ensure the system operates at peak efficiency and reliability. When a minor issue is detected through the monitoring system, the facility's trained maintenance staff quickly diagnoses and resolves the problem, minimizing any disruption to the hospital's operations.

By prioritizing proper installation, maintenance, and monitoring of your commercial geothermal system, you can ensure its long-term performance, efficiency, and reliability while maximizing the benefits of this clean, sustainable energy source. Remember that investing in quality installation, regular maintenance, and ongoing monitoring can help you avoid costly repairs, minimize downtime, and optimize your geothermal system's contributions to your business's energy needs and sustainability goals.

As you embark on your journey to harness geothermal energy for your business, keep in mind the importance of careful planning, design, installation, and ongoing management of your geothermal system. By working with experienced professionals and dedicating the necessary resources to these critical aspects, you can unlock the full potential of geothermal energy to power your business while reducing your environmental impact and operating costs.

Part III: Implementing Renewable Energy Solutions for Businesses

Chapter 8
Assessing Your Business's Energy Needs

Conducting a Commercial Energy Audit

Before implementing renewable energy solutions for your business, it's essential to have a clear understanding of your company's energy needs and consumption patterns. This information will help you determine the most appropriate and cost-effective renewable energy technologies for your business. In this chapter, we'll focus on assessing your business's energy needs through the process of conducting a commercial energy audit.

Conducting a Commercial Energy Audit

A commercial energy audit is a comprehensive assessment of your business's energy usage, efficiency, and potential for improvement. By conducting an energy audit, you can identify areas of energy waste, inefficiencies, and opportunities for incorporating renewable energy technologies. Here's a step-by-step guide to conducting a commercial energy audit:

Step 1: Gather and Analyze Energy Bills
Begin by gathering your business's energy bills for the past year, including electricity, natural gas, and any other fuels used.

Analyze these bills to determine your total energy consumption, peak demand, and seasonal variations. This data will serve as a baseline for identifying potential energy savings and renewable energy opportunities.

Step 2: Inventory Energy-Consuming Equipment and Systems
Create a comprehensive inventory of all energy-consuming equipment and systems in your business, including:

- HVAC (Heating, Ventilation, and Air Conditioning) systems
- Lighting fixtures and controls
- Office equipment (computers, printers, copiers, etc.)
- Production machinery and equipment
- Refrigeration and food service equipment
- Water heating systems

For each item, note its age, efficiency rating, and hours of operation. This inventory will help you identify inefficient equipment that may need to be upgraded or replaced.

Step 3: Conduct a Building Envelope Assessment
Assess your building's envelope, which includes the walls, roof, windows, and doors, to identify any areas of energy loss. Look for gaps, cracks, or inadequate insulation that may be allowing conditioned air to escape or outside air to infiltrate. Consider the age and condition of your windows and doors, as well as the insulation levels in your walls and roof.

Step 4: Evaluate Lighting Systems
Examine your business's lighting systems, including indoor and outdoor lighting, to identify opportunities for energy savings. Consider the type of light bulbs used (incandescent, fluorescent, LED), the age and condition of fixtures, and the use of lighting controls such as occupancy sensors or dimmers. Identify areas where natural light can be utilized to reduce the need for artificial lighting.

Step 5: Assess HVAC Systems
Evaluate the efficiency and condition of your HVAC systems, which can account for a significant portion of your business's energy consumption. Consider the age and efficiency ratings of your heating and cooling equipment, as well as the condition of ductwork and insulation. Look for opportunities to improve temperature controls, such as programmable thermostats or zoning systems.

Step 6: Monitor and Analyze Energy Consumption
Implement a system for monitoring and analyzing your business's energy consumption on an ongoing basis. This may involve installing submeters to track energy usage by department, process, or equipment. Regularly review your energy data to identify trends, anomalies, or areas for improvement.

Step 7: Consider Renewable Energy Opportunities
Based on your energy audit findings, consider potential renewable energy opportunities for your business. This may include:

- Solar PV systems for electricity generation
- Solar thermal systems for water heating or process heat
- Geothermal heat pumps for efficient heating and cooling
- Wind turbines for electricity generation
- Biomass systems for heating or combined heat and power (CHP)

Evaluate the feasibility and cost-effectiveness of each option based on your business's specific energy needs, location, and available resources.

Imagine a manufacturing facility that conducts a comprehensive energy audit. Through the audit process, they discover that their aging HVAC system is operating inefficiently, leading to significant energy waste. They also identify opportunities to upgrade to LED lighting and install occupancy sensors in low-traffic areas. Based on their energy consumption data and the audit findings, the facility managers determine that a rooftop solar PV system could offset a substantial portion of their electricity needs, providing long-term cost savings and environmental benefits. By conducting a thorough energy audit, the manufacturing facility is able to make informed decisions about implementing renewable energy solutions that align with their specific needs and goals.

Conducting a commercial energy audit is a crucial step in assessing your business's energy needs and identifying opportunities for renewable energy integration.

By following the steps outlined above and engaging the help of experienced energy professionals, you can develop a comprehensive understanding of your company's energy profile and make data-driven decisions about implementing renewable energy solutions.

In the next section, we'll explore how to determine your business's renewable energy potential based on the findings of your energy audit and other key factors, such as location, available space, and local renewable energy resources.

Determining Your Renewable Energy Potential

Once you have conducted a thorough energy audit and have a clear understanding of your business's energy needs and consumption patterns, the next step is to determine your renewable energy potential. This process involves assessing the feasibility and viability of implementing various renewable energy technologies based on factors such as your location, available space, local resources, and economic considerations. In this section, we'll explore the key steps in determining your business's renewable energy potential.

Step 1: Evaluate Your Location and Climate
Your business's location and climate play a crucial role in determining which renewable energy technologies are most suitable. Consider the following factors:

- Solar Resources: Assess your location's solar potential by evaluating factors such as average daily sunlight hours, solar irradiance levels, and shading from nearby buildings or trees. Regions with abundant, consistent sunlight are ideal for solar PV and solar thermal systems.
- Wind Resources: Determine your area's average wind speeds and patterns using wind resource maps or local meteorological data. Consistent, strong winds are necessary for the effective operation of wind turbines.
- Geothermal Resources: Consider the availability and accessibility of geothermal resources in your region, such as hot springs, geysers, or high-temperature aquifers. Proximity to these resources can make geothermal energy systems more viable.

- Biomass Resources: Evaluate the availability of local biomass resources, such as wood waste, agricultural residues, or municipal solid waste, which can be used as fuel for biomass heating or combined heat and power (CHP) systems.

Step 2: Assess Available Space and Infrastructure

Consider the physical space and infrastructure available at your business site for implementing renewable energy technologies:

- Rooftop Space: Determine the size, orientation, and condition of your rooftop for potential solar PV or solar thermal installations. Large, unobstructed, south-facing roofs (in the northern hemisphere) are ideal for maximizing solar energy production.
- Land Area: Assess the availability and suitability of land on your business property for ground-mounted solar arrays, wind turbines, or geothermal systems. Consider factors such as land use restrictions, soil conditions, and proximity to buildings or infrastructure.
- Building Infrastructure: Evaluate your building's structural integrity, electrical systems, and plumbing to determine the feasibility of integrating renewable energy technologies. Some technologies, such as solar PV or geothermal heat pumps, may require upgrades or modifications to your existing infrastructure.

Step 3: Analyze Energy Load Profiles

Use the data from your energy audit to analyze your business's energy load profiles, which show how your energy consumption varies over time (daily, weekly, seasonally).

This information can help you determine the optimal size and configuration of renewable energy systems to meet your specific energy needs.

- Peak Demand: Identify periods of peak energy demand and consider how renewable energy technologies can help reduce or shift these peaks to minimize stress on the grid and potentially lower demand charges.
- Baseload Consumption: Determine your business's consistent, baseline energy needs and consider renewable energy options that can provide reliable, steady power, such as geothermal or biomass systems.
- Seasonal Variations: Assess how your energy consumption varies across seasons and consider renewable energy technologies that can complement these variations. For example, solar PV may provide more energy during summer months, while wind turbines may perform better during winter.

Step 4: Evaluate Economic Feasibility

Assess the economic feasibility of implementing renewable energy technologies by considering factors such as upfront costs, long-term savings, incentives, and financing options.

- Cost-Benefit Analysis: Conduct a cost-benefit analysis to compare the upfront costs of renewable energy systems with the projected long-term energy savings and other benefits, such as reduced greenhouse gas emissions or improved energy security.

- Incentives and Rebates: Research available government incentives, tax credits, and rebates for businesses investing in renewable energy technologies. These incentives can significantly reduce the upfront costs and improve the financial viability of your projects.
- Financing Options: Explore various financing options, such as loans, leases, or power purchase agreements (PPAs), which can help spread the upfront costs of renewable energy systems over time and align with your business's cash flow and budget constraints.

Step 5: Consult with Renewable Energy Professionals

Partner with experienced renewable energy professionals, such as consultants, engineers, or installers, who can provide expert guidance and support in determining your business's renewable energy potential. These professionals can help you:

- Conduct detailed site assessments and feasibility studies
- Identify the most suitable renewable energy technologies for your specific needs and constraints
- Optimize system designs and configurations for maximum performance and cost-effectiveness
- Navigate regulatory requirements, permitting processes, and interconnection procedures
- Develop comprehensive project proposals and financial models to support decision-making

Imagine a food processing company that has completed an energy audit and is now evaluating its renewable energy potential. The company's facilities are located in an area with abundant solar resources and ample rooftop space for solar PV installations. By analyzing their energy load profiles, the company identifies significant daytime electricity consumption that aligns well with solar PV generation. They also discover that their consistent hot water needs could be met by a solar thermal system. Through a cost-benefit analysis and consultation with renewable energy professionals, the company determines that a combination of solar PV and solar thermal technologies would provide substantial long-term energy savings and environmental benefits, with a reasonable payback period. This information enables the company to make an informed decision to pursue these renewable energy projects.

Determining your business's renewable energy potential is a multi-faceted process that requires careful consideration of various factors, from location and climate to economic feasibility and expert consultation. By thoroughly assessing your renewable energy options and aligning them with your business's specific needs and goals, you can make informed decisions about implementing sustainable energy solutions that provide long-term benefits for your company and the environment.

Chapter 9
Designing a Solar Panel System for Your Business

Sizing Your Commercial Solar Panel Array

After determining that your business has good potential for solar energy generation, the next step is to design a solar panel system that meets your company's energy needs, aligns with your available space and infrastructure, and optimizes your return on investment. Designing a commercial solar panel system involves several key considerations, such as sizing your solar panel array, selecting the right components, and planning for installation and maintenance. In this chapter, we'll focus on sizing your commercial solar panel array.

Sizing Your Commercial Solar Panel Array
Properly sizing your commercial solar panel array is crucial for ensuring that your system generates enough electricity to meet your business's energy needs while optimizing your investment and available space. Here's a step-by-step guide to sizing your commercial solar panel array:

Step 1: Determine Your Energy Consumption
Begin by analyzing your business's energy consumption data from your energy audit. Determine your average daily, monthly, and annual electricity consumption in kilowatt-hours (kWh). This information will serve as the basis for sizing your solar panel array.

Step 2: Assess Your Available Space
Evaluate the available space for installing solar panels at your business site. This may include rooftops, carports, or ground-mounted areas. Measure the dimensions of these areas and consider any obstructions, such as HVAC units, skylights, or shading from nearby buildings or trees. This will help you determine the maximum number of solar panels that can be installed.

Step 3: Evaluate Solar Potential
Assess your location's solar potential by considering factors such as:

- Solar Irradiance: Determine the average daily solar irradiance (the amount of solar energy that reaches a given area) for your location. This is typically measured in kilowatt-hours per square meter per day (kWh/m2/day).
- Sun Hours: Estimate the average number of peak sun hours per day for your location. Peak sun hours refer to the number of hours during which the solar irradiance equals 1 kW/m2.
- Shading and Orientation: Consider the orientation of your available space (ideally south-facing in the northern hemisphere) and any potential shading from nearby objects, as these factors can impact the efficiency of your solar panels.

Step 4: Select Your Solar Panels
Choose the type and efficiency of solar panels that best suit your needs and budget. Factors to consider include:

- Panel Type: Decide between monocrystalline, polycrystalline, or thin-film solar panels based on their efficiency, cost, and space requirements.
- Panel Wattage: Consider the wattage of individual solar panels, which typically ranges from 250 to 400 watts per panel. Higher wattage panels can generate more electricity in a smaller space but may also be more expensive.
- Panel Efficiency: Evaluate the efficiency of different solar panel options, which represents the percentage of sunlight that is converted into electricity. Higher efficiency panels can generate more electricity in a given space but may come at a higher cost.

Step 5: Calculate Your Array Size

To determine the size of your solar panel array, use the following formula:

Array Size (kW) = Daily Energy Consumption (kWh) ÷ (Solar Irradiance (kWh/m2/day) × Panel Efficiency)

For example, if your business consumes 500 kWh of electricity per day, your location receives an average solar irradiance of 5 kWh/m2/day, and you choose solar panels with an efficiency of 18%, your array size would be:

Array Size (kW) = 500 kWh ÷ (5 kWh/m2/day × 0.18) = 55.56 kW

To determine the number of panels needed, divide the array size by the wattage of your chosen panels. For instance, if you select 300-watt panels, you would need:

Number of Panels = 55.56 kW ÷ 0.3 kW = 185.2 (rounded up to 186 panels)

Step 6: Optimize Your Array Layout

Based on your available space and the number of panels needed, design an optimal layout for your solar panel array. Consider factors such as:

- Tilt Angle: Determine the ideal tilt angle for your solar panels based on your latitude and the seasonal variations in sun position. This will help maximize solar energy capture.
- Row Spacing: Ensure adequate spacing between rows of panels to minimize shading and maintain accessibility for maintenance.
- Mounting Systems: Select a suitable mounting system (roof-mounted, ground-mounted, or carport) that secures your panels while allowing for proper ventilation and cable management.

Step 7: Assess Energy Storage Options

Consider incorporating energy storage solutions, such as batteries, into your solar panel system design. Energy storage can help you:

- Manage Peak Demand: Store excess solar energy generated during peak sunlight hours for use during periods of high energy demand or when solar production is low.
- Increase Energy Resilience: Provide backup power during grid outages or emergencies, ensuring business continuity and reducing downtime.

- Optimize Energy Costs: Take advantage of time-of-use (TOU) electricity rates by storing solar energy during off-peak hours and using it during peak periods when rates are higher.

Imagine a manufacturing facility that consumes an average of 2,000 kWh of electricity per day. The facility has a large, unshaded rooftop area that can accommodate up to 1,000 solar panels. The location receives an average solar irradiance of 4.5 kWh/m2/day. The facility managers decide to use monocrystalline solar panels with an efficiency of 20% and a wattage of 350 watts per panel. Using the formula above, they calculate that they need a 222.22 kW solar panel array, which equates to 635 panels (rounded up from 634.92). They design an optimized layout that maximizes energy production while ensuring adequate spacing for maintenance and ventilation. To further enhance their energy resilience and manage peak demand, they incorporate a lithium-ion battery storage system sized to provide backup power for critical loads during grid outages.

Properly sizing your commercial solar panel array is essential for maximizing the benefits of your solar investment while ensuring that your system meets your business's energy needs and aligns with your available space and budget. By following the steps outlined above and working with experienced solar professionals, you can design a solar panel system that optimizes your energy production, cost savings, and environmental impact.

In the next section, we'll explore how to choose the right components for your commercial solar panel system, including inverters, mounting hardware, and monitoring systems, to ensure optimal performance, reliability, and longevity.

Choosing the Right Components

Once you have determined the size of your commercial solar panel array, the next step is to select the right components to ensure optimal system performance, reliability, and safety. The main components of a commercial solar panel system include solar panels, inverters, mounting hardware, and monitoring systems. In this section, we'll discuss how to choose the right components for your business's solar panel system.

1. Solar Panels

In addition to the factors discussed in the previous section (panel type, wattage, and efficiency), consider the following when selecting solar panels for your commercial system:

- Durability: Choose solar panels with robust construction and high-quality materials that can withstand harsh weather conditions, such as high winds, hail, or extreme temperatures. Look for panels with strong frames, tempered glass, and reliable sealing to ensure long-term durability.
- Warranty: Opt for solar panels with comprehensive warranties that cover both product defects and performance. Look for panels with product warranties of at least 10-12 years and performance warranties that guarantee a certain level of output (usually 80-90%) over 25 years or more.
- Certifications: Ensure that your chosen solar panels meet relevant safety and quality standards, such as UL 1703 or IEC 61215, and are certified by reputable organizations like the International Electrotechnical Commission (IEC) or the California Energy Commission (CEC).

2. Inverters

Inverters convert the direct current (DC) electricity generated by your solar panels into alternating current (AC) electricity that can be used by your business or exported to the grid. When selecting inverters for your commercial solar panel system, consider:

- Type: There are three main types of inverters: string inverters, microinverters, and power optimizers. String inverters are the most common and cost-effective option for commercial systems, while microinverters and power optimizers offer module-level optimization and monitoring capabilities.
- Efficiency: Choose inverters with high efficiency ratings (typically above 95%) to minimize energy losses during the DC to AC conversion process.
- Capacity: Ensure that your inverters have sufficient capacity to handle the maximum output of your solar panel array. Consider factors such as the number of panels per string and the total system size when selecting inverter capacity.
- Warranty: Look for inverters with comprehensive warranties that cover both product defects and performance. Inverter warranties typically range from 5-15 years, with some manufacturers offering extended warranties for an additional cost.

3. Mounting Hardware

Mounting hardware secures your solar panels to your business's roof, carport, or ground-mounted structure. When selecting mounting hardware, consider:

- Compatibility: Choose mounting hardware that is compatible with your specific solar panels and installation site. Ensure that the hardware is designed to withstand the weight and dimensions of your panels and can accommodate your roof type or ground conditions.
- Durability: Opt for mounting hardware made from high-quality, corrosion-resistant materials, such as aluminum or stainless steel, to ensure long-term durability and structural integrity.
- Adjustability: Consider mounting hardware that allows for adjustable tilt angles to optimize solar energy capture based on your location and seasonal variations in sun position.
- Warranty: Look for mounting hardware with warranties that cover defects and structural integrity for at least 10-25 years.

4. Monitoring Systems

Monitoring systems allow you to track your solar panel system's performance, energy production, and any potential issues in real-time. When selecting a monitoring system, consider:

- Data Accuracy: Choose a monitoring system that provides accurate, real-time data on your system's performance, including energy production, consumption, and environmental factors like temperature and irradiance.
- User Interface: Opt for a monitoring system with a user-friendly interface that allows you to easily access and interpret your system's data, either through a web portal or mobile app.

- Alerts and Diagnostics: Look for monitoring systems that offer automated alerts and diagnostic tools to help you quickly identify and address any performance issues or system faults.
- Integration: Consider monitoring systems that can integrate with your existing building management or energy management systems, allowing for seamless data exchange and analysis.

Imagine a retail store that has designed a 100 kW rooftop solar panel system. They select high-efficiency monocrystalline panels with a 25-year performance warranty and pair them with a string inverter that offers 98% efficiency and a 10-year warranty. For mounting hardware, they choose a rail-based system with adjustable tilt angles, made from anodized aluminum for long-term durability. To monitor their system's performance, they opt for a cloud-based monitoring platform that provides real-time data on energy production, consumption, and environmental factors, accessible through both a web portal and a mobile app. The monitoring system also integrates with their existing building management system, allowing for comprehensive energy analysis and optimization.

By carefully selecting the right components for your commercial solar panel system, you can ensure optimal performance, reliability, and safety while maximizing your return on investment. Consider factors such as durability, efficiency, compatibility, and warranties when choosing solar panels, inverters, mounting hardware, and monitoring systems. Working with experienced solar professionals can help you navigate the selection process and identify the best components for your specific business needs and installation site.

In the next section, we'll discuss the installation, commissioning, and maintenance considerations for your commercial solar panel system, including best practices for ensuring a smooth and successful installation process and ongoing system performance.

Installation, Maintenance, and Monitoring

Once you have designed your commercial wind turbine system and selected the appropriate components, the next critical steps are proper installation, maintenance, and monitoring. These aspects ensure that your wind turbine operates safely, efficiently, and reliably over its expected lifespan. In this section, we'll discuss the key considerations for the installation, maintenance, and monitoring of your commercial wind turbine system.

Installation

1. Site Preparation

Before installing your wind turbine, ensure that the site is properly prepared. This includes:

- Clearing the area of any obstacles or vegetation that could interfere with the turbine's operation
- Grading and leveling the site to ensure a stable foundation for the turbine tower
- Constructing access roads or paths for installation and maintenance equipment

2. Foundation Construction

A sturdy foundation is critical for the stability and longevity of your wind turbine. The foundation design will depend on factors such as the turbine size, soil conditions, and local building codes. Common foundation types include:

- Concrete slab: A reinforced concrete slab that supports the turbine tower and anchors it to the ground

- Pier and anchor bolt: A deep concrete pier with anchor bolts that secures the tower to the foundation
- Gravity foundation: A large concrete mass that uses its weight to provide stability for the turbine tower

3. Tower and Turbine Erection
Once the foundation is in place, the tower and turbine can be erected. This typically involves:

- Assembling the tower sections and lifting them into place using a crane
- Securing the tower to the foundation using anchor bolts or other fastening methods
- Installing the turbine nacelle, rotor, and blades at the top of the tower
- Connecting the turbine to the electrical and control systems

4. Electrical and Control System Integration
The wind turbine must be properly integrated with the electrical grid and control systems. This involves:

- Installing transformers and switchgear to convert the turbine's output to grid-compatible voltage and frequency
- Connecting the turbine to the grid through underground cables or overhead power lines
- Installing control systems and sensors to monitor and regulate the turbine's operation
- Implementing safety features such as overspeed protection, emergency brakes, and lightning protection

Maintenance

1. Regular Inspections
To ensure the ongoing performance and safety of your wind turbine, conduct regular inspections at least once a year, or more frequently if recommended by the manufacturer. During these inspections:

- Check the turbine blades, nacelle, and tower for any signs of damage, wear, or corrosion
- Inspect the electrical and control systems for proper functioning and any signs of damage or deterioration
- Verify that all bolts, fasteners, and structural components are secure and properly torqued
- Check for any leaks or damage to the lubrication and cooling systems

2. Lubrication and Fluid Replacement
Wind turbines rely on proper lubrication and fluid levels to operate efficiently and prevent premature wear. Regularly:

- Grease bearings and gears according to the manufacturer's recommendations
- Change oil and other fluids in the gearbox, hydraulic systems, and other components as specified
- Check for any leaks or contamination in the lubrication and cooling systems

3. Blade and Rotor Maintenance
The turbine blades and rotor are exposed to significant stresses and environmental factors. To maintain their performance and integrity:

- Inspect the blades for any cracks, delamination, or other damage
- Clean the blades to remove any dirt, debris, or ice accumulation that could affect their aerodynamics
- Balance the rotor and check for any vibrations or oscillations that could indicate an imbalance or structural issue

4. Electrical and Control System Maintenance
Regularly maintain and test the electrical and control systems to ensure their proper functioning and safety. This includes:

- Inspecting transformers, switchgear, and other electrical components for any signs of damage or wear
- Testing the turbine's safety features, such as emergency brakes and overspeed protection, to verify their operation
- Updating control system software and firmware as needed to optimize performance and address any known issues

Monitoring:

1. SCADA Systems
Supervisory Control and Data Acquisition (SCADA) systems are used to monitor and control wind turbines remotely. These systems:

- Collect real-time data on the turbine's performance, including power output, wind speed, and component temperatures
- Alert operators to any faults or abnormal conditions that require attention

- Allow for remote control and adjustment of the turbine's operation

2. Performance Analysis
Regularly analyze the data collected by the SCADA system to:

- Track the turbine's energy production and efficiency over time
- Identify any trends or patterns that may indicate underlying issues or opportunities for optimization
- Compare the turbine's performance to expected or modeled output based on wind resource data

3. Predictive Maintenance
Use the data collected by the SCADA system and other condition monitoring systems to develop predictive maintenance strategies. This involves:

- Analyzing data on component vibration, temperature, and other parameters to identify potential failure modes
- Scheduling maintenance or component replacements based on predictive models, rather than fixed intervals
- Optimizing maintenance resources and minimizing downtime by addressing issues before they cause failures

Imagine a rural municipality that has installed a 2 MW wind turbine to provide clean energy for its residents. The installation team carefully prepares the site, constructs a robust concrete foundation, and erects the tower and turbine using a specialized crane. The turbine is then connected to the electrical grid and control systems, with comprehensive safety features and monitoring capabilities. The municipality contracts with a specialized wind turbine maintenance company to perform regular inspections, lubrication, and repairs, while also monitoring the turbine's performance through a SCADA system. When the data analysis reveals a potential issue with the gearbox, the maintenance team schedules a proactive replacement, minimizing downtime and preventing a more costly failure.

By prioritizing proper installation, maintenance, and monitoring of your commercial wind turbine system, you can optimize its performance, safety, and longevity. Regular inspections, preventive maintenance, and data-driven monitoring help you maximize energy production, minimize downtime, and extend the lifespan of your investment. As with any complex system, working with experienced professionals and establishing a comprehensive maintenance and monitoring plan is essential for the long-term success of your wind energy project.

As you navigate the process of harnessing wind energy for your business, keep in mind the importance of site selection, system design, component quality, and ongoing care. By following best practices and partnering with knowledgeable experts, you can unlock the full potential of wind power to support your business's energy needs, sustainability goals, and financial objectives.

Chapter 10
Designing a Wind Turbine System for Your Business

Assessing Wind Resources at Your Business Location

After determining that your business has good potential for wind energy generation, the next step is to design a wind turbine system that meets your company's energy needs, aligns with your available space and resources, and optimizes your return on investment. Designing a commercial wind turbine system involves several key considerations, such as assessing your wind resources, selecting the appropriate turbine size and type, and planning for installation and maintenance. In this chapter, we'll focus on assessing wind resources at your business location.

Assessing Wind Resources at Your Business Location
A thorough assessment of the wind resources at your business location is essential for determining the feasibility and potential performance of a wind turbine system. This process involves evaluating factors such as wind speed, direction, and consistency, as well as considering site-specific characteristics that may impact turbine installation and operation. Here's a step-by-step guide to assessing wind resources at your business location:

Step 1: Consult Wind Resource Maps

Begin by consulting wind resource maps, which provide estimated average wind speeds and power densities for different regions. These maps can give you a general idea of the wind potential in your area. Some useful resources include:

- National Renewable Energy Laboratory (NREL) Wind Resource Maps
- Global Wind Atlas
- State or local wind resource maps

Keep in mind that these maps provide a high-level overview and may not capture site-specific variations in wind speed and direction.

Step 2: Collect Local Wind Data

To get a more accurate picture of your site's wind resources, collect local wind data from nearby weather stations, airports, or other wind monitoring sites. Look for data on:

- Average wind speeds at different heights
- Wind speed distribution (how often the wind blows at different speeds)
- Prevailing wind directions
- Seasonal and diurnal (day vs. night) wind patterns

This data can help you understand the general wind characteristics in your area and identify any potential challenges or opportunities for wind energy development.

Step 3: Conduct On-Site Wind Measurements
For the most accurate assessment of your site's wind resources, conduct on-site wind measurements using specialized equipment such as meteorological towers or remote sensing devices (e.g., sodar or lidar). This typically involves:

- Installing wind measurement equipment at the proposed turbine location
- Collecting data on wind speed, direction, and turbulence at multiple heights (e.g., 40, 60, and 80 meters) for at least one year
- Analyzing the data to determine the average wind speed, wind shear (how wind speed changes with height), and turbulence intensity
- Extrapolating the data to the proposed turbine hub height using established mathematical models

On-site measurements provide the most reliable data for estimating your site's wind energy potential and informing turbine selection and system design.

Step 4: Evaluate Site-Specific Characteristics
In addition to wind speed and direction, consider site-specific characteristics that may impact the feasibility and performance of a wind turbine system. These factors include:

- Terrain and Topography: Assess the site's elevation, slope, and surrounding terrain features that may affect wind flow patterns and turbine installation.

- Surface Roughness: Evaluate the site's surface characteristics, such as vegetation, buildings, or other obstacles, which can create turbulence and reduce wind speeds.
- Accessibility: Consider the site's accessibility for construction equipment, maintenance vehicles, and electrical infrastructure.
- Environmental and Permitting Constraints: Identify any environmental sensitivities, such as wildlife habitats or migration routes, and permitting requirements that may affect turbine siting and operation.

Step 5: Estimate Energy Production Potential

Based on the collected wind data and site-specific characteristics, estimate the potential energy production of a wind turbine system at your business location. This typically involves:

- Selecting a representative wind turbine model with a power curve that matches your site's wind speed distribution
- Calculating the expected annual energy production (AEP) using the turbine's power curve and your site's wind speed frequency distribution
- Applying any necessary adjustments for factors such as wind shear, turbulence, and array losses (if installing multiple turbines)
- Comparing the estimated energy production to your business's energy needs and financial goals

This energy production estimate will help you determine the economic feasibility of a wind turbine system and inform subsequent design and planning decisions.

Imagine a manufacturing facility located on a hilltop in a rural area. They consult wind resource maps and find that their region has an estimated average wind speed of 6.5 meters per second at 80 meters height, indicating good potential for wind energy development. They then collect wind data from a nearby airport and find that the prevailing winds are from the southwest, with higher speeds during the winter months. To further refine their assessment, they install a meteorological tower and collect on-site wind data for a full year. The data confirms an average wind speed of 7.0 meters per second at 80 meters, with minimal turbulence and wind shear. Based on this data and the site's favorable terrain and accessibility, they estimate that a 1.5 MW wind turbine could generate approximately 4,500 MWh of electricity per year, meeting a significant portion of their energy needs.

By thoroughly assessing the wind resources at your business location, you can make informed decisions about the feasibility and potential benefits of a wind turbine system. This assessment lays the foundation for subsequent steps in the design process, such as turbine selection and siting, financial modeling, and permitting.

In the next section, we'll explore how to select the appropriate wind turbine for your business based on your wind resource assessment, energy needs, and site-specific constraints.

Choosing the Right Commercial Wind Turbine

Once you have assessed the wind resources at your business location and determined that a wind turbine system is feasible, the next step is to choose the right commercial wind turbine for your needs. Selecting the appropriate turbine involves considering factors such as turbine size, type, efficiency, and compatibility with your site's specific wind and environmental conditions. In this section, we'll discuss how to choose the right commercial wind turbine for your business.

1. Determine Turbine Size and Capacity

The size and capacity of your wind turbine will depend on your business's energy needs, wind resource assessment, and available space. Commercial wind turbines range from small, distributed systems (under 100 kW) to large, utility-scale turbines (1 MW or more). When selecting turbine size and capacity, consider:

- Energy Production: Choose a turbine size that can generate enough energy to meet a significant portion of your business's energy needs, based on your wind resource assessment and estimated annual energy production.

- Site Constraints: Ensure that your chosen turbine size is compatible with your site's available space, setback requirements, and any height restrictions.

- Economies of Scale: Generally, larger turbines offer better economies of scale, with lower costs per kilowatt-hour of energy produced. However, they also require more space and higher upfront investments.

2. Evaluate Turbine Type and Design

There are several types of wind turbines, each with its own advantages and suitability for different sites and applications. The most common types include:

- Horizontal Axis Wind Turbines (HAWTs): HAWTs are the most common type of commercial wind turbine, with the rotor shaft and blades positioned horizontally. They are typically more efficient and cost-effective than other designs, but require consistent, unidirectional wind flow.

- Vertical Axis Wind Turbines (VAWTs): VAWTs have the rotor shaft positioned vertically, with the blades rotating around it. They are less common than HAWTs but can be suitable for sites with more turbulent or variable wind conditions. However, they are generally less efficient and have lower energy production.

- Ducted or Shrouded Turbines: These turbines feature a shroud or duct around the blades, which can increase wind speed and energy production in some conditions. They may be suitable for sites with lower wind speeds or space constraints, but are generally more expensive and less proven than traditional designs.

3. Compare Turbine Efficiency and Performance

When choosing a commercial wind turbine, compare the efficiency and performance of different models to ensure you select the best option for your site and needs. Key factors to consider include:

- Power Curve: The power curve shows how much power the turbine generates at different wind speeds. Look for a turbine with a power curve that matches your site's specific wind speed distribution for optimal energy production.

- Capacity Factor: The capacity factor is the ratio of the turbine's actual energy production to its theoretical maximum output. Higher capacity factors indicate more efficient and cost-effective turbines.

- Cut-In and Cut-Out Speeds: The cut-in speed is the minimum wind speed at which the turbine starts generating power, while the cut-out speed is the maximum wind speed at which the turbine shuts down to prevent damage. Ensure that your chosen turbine's cut-in and cut-out speeds are compatible with your site's wind conditions.

4. Assess Durability and Reliability

Commercial wind turbines are significant investments that must withstand harsh environmental conditions and operate reliably for decades. When selecting a turbine, consider:

- Proven Track Record: Choose turbines from established manufacturers with a proven track record of reliability and performance in real-world conditions.

- Material Quality: Look for turbines made with high-quality, durable materials that can withstand the stresses of constant operation and exposure to the elements.

- Warranty and Service: Consider the manufacturer's warranty terms and the availability of local service and support for maintenance and repairs.

5. Evaluate Compatibility and Integration

Finally, ensure that your chosen wind turbine is compatible with your site's electrical infrastructure and can be seamlessly integrated with your business's energy systems. This may involve:

- Electrical Compatibility: Verify that the turbine's output voltage and frequency are compatible with your site's electrical grid and any existing energy systems.

- Interconnection Requirements: Understand the requirements and process for interconnecting the turbine to your local utility grid, including any necessary upgrades or modifications to your electrical infrastructure.

- Energy Storage and Management: Consider integrating energy storage systems, such as batteries, to store excess energy production and provide consistent power during low-wind periods. Also, evaluate energy management systems that can optimize the turbine's operation and integrate with your business's overall energy strategy.

Imagine a rural agricultural processing facility that has determined through a wind resource assessment that a 1 MW wind turbine would be suitable for their energy needs and site conditions. They compare several HAWT models from reputable manufacturers and select a turbine with a power curve that closely matches their site's wind speed distribution, a high capacity factor, and a proven track record of reliability in similar environments. The chosen turbine is also compatible with their existing electrical infrastructure and can be easily interconnected to the local utility grid. To maximize the value of their investment, they also integrate a battery storage system to store excess energy production and provide consistent power for their critical processing equipment.

By carefully selecting the right commercial wind turbine based on your business's energy needs, site conditions, and performance requirements, you can ensure that your wind energy investment delivers reliable, cost-effective, and sustainable power for years to come. Remember to work with experienced wind energy professionals who can guide you through the selection process and help you make informed decisions based on your unique circumstances.

Installation, Maintenance, and Monitoring

After selecting the right commercial geothermal system for your business, the next crucial steps are proper installation, regular maintenance, and ongoing monitoring to ensure optimal performance, efficiency, and longevity. Installing a geothermal system involves several key considerations, such as site preparation, drilling, and system integration, while regular maintenance and monitoring are essential for identifying and addressing any issues that may arise. In this section, we'll discuss the installation, maintenance, and monitoring requirements for your commercial geothermal system.

Installation

1. Site Preparation and Permitting
Before installing your geothermal system, properly prepare the site and obtain all necessary permits and approvals. This includes:

- Conducting a detailed site survey to identify any potential obstacles or constraints, such as underground utilities or environmental sensitivities.
- Obtaining all required permits and approvals from local, state, and federal authorities, such as building permits, drilling permits, and environmental assessments.
- Preparing the site for drilling and installation, including clearing and grading the area, establishing access roads, and setting up any necessary safety measures.

2. Drilling and Well Installation

The drilling and installation of geothermal wells is a critical component of the installation process. This involves:

- Drilling the geothermal wells to the required depth and diameter based on the system design and geological conditions. This may involve using specialized drilling rigs and techniques, such as directional drilling or casing installation.
- Installing the well casing, grouting, and any necessary well screens or packers to ensure the integrity and longevity of the well.
- Conducting well logging and testing to verify the well's integrity, productivity, and thermal properties.

3. Heat Exchanger and Piping Installation

After the wells are installed, the heat exchanger and piping system must be installed to transfer heat between the geothermal fluid and the building's HVAC system. This includes:

- Installing the heat exchanger, circulation pumps, and any necessary valves or controls based on the system design and manufacturer's specifications.
- Connecting the heat exchanger to the geothermal well piping and the building's HVAC system, ensuring proper insulation and sealing to minimize heat loss.
- Pressure testing and flushing the piping system to verify its integrity and remove any debris or contaminants.

4. System Integration and Commissioning
The final step in the installation process is integrating the geothermal system with the building's existing HVAC and electrical systems and commissioning the system for operation. This involves:

- Connecting the geothermal system to the building's HVAC system, including any necessary modifications or upgrades to the existing ductwork, piping, or control systems.
- Installing and programming the system controls, monitoring equipment, and any necessary safety devices or alarms.
- Conducting a thorough commissioning process to verify that the system is operating as designed, including testing and balancing the system, calibrating sensors and controls, and providing operator training.

Maintenance and Monitoring

1. Regular Inspection and Testing
To ensure the ongoing performance and efficiency of your commercial geothermal system, conduct regular inspections and testing, including:

- Visual inspections of the wells, piping, heat exchanger, and other system components to identify any signs of wear, damage, or leakage.
- Pressure and flow testing to verify that the system is operating within the designed parameters and to detect any blockages or restrictions.

- Water quality testing to monitor the geothermal fluid's chemistry and to identify any potential scaling, corrosion, or biological growth issues.

2. Preventive Maintenance

Implement a comprehensive preventive maintenance program to proactively address any potential issues and extend the lifespan of your geothermal system. This may include:

- Regularly cleaning and flushing the heat exchanger and piping system to remove any accumulated debris or scale.
- Lubricating and replacing any worn or damaged components, such as pumps, valves, or bearings, based on the manufacturer's recommendations.
- Updating and calibrating the system controls and monitoring equipment to ensure optimal performance and efficiency.

3. Performance Monitoring and Optimization

Continuously monitor and optimize your geothermal system's performance to maximize energy savings and minimize operating costs. This involves:

- Installing and utilizing a comprehensive monitoring system to track key performance indicators, such as energy consumption, system efficiency, and operating temperatures.
- Analyzing performance data to identify any trends, anomalies, or opportunities for improvement, such as adjusting system setpoints or reconfiguring piping layouts.

- Implementing any necessary system modifications or upgrades to optimize performance and efficiency based on the monitoring data and analysis.

4. Emergency Response and Troubleshooting

Establish a clear protocol for responding to any system emergencies or malfunctions and troubleshooting any issues that arise. This includes:

- Developing a detailed emergency response plan that outlines the procedures for shutting down the system, isolating any damaged components, and notifying relevant personnel.
- Training facility staff and maintenance personnel on the proper operation and troubleshooting of the geothermal system, including identifying and diagnosing common issues.
- Establishing a network of qualified geothermal service providers and equipment suppliers to quickly respond to any emergencies or repair needs.

Imagine a hospital that has installed a large-scale ground source heat pump system to provide heating and cooling for its facilities. The installation process involved careful site preparation, including a detailed site survey and obtaining all necessary permits and approvals. The geothermal wells were drilled to a depth of 500 feet using specialized directional drilling techniques, and the well casings were installed and grouted to ensure long-term integrity. The heat exchanger and piping system were installed and integrated with the hospital's existing HVAC system, and the entire system was thoroughly commissioned to verify optimal performance.

The hospital implements a comprehensive maintenance and monitoring program, including regular inspections, preventive maintenance, and continuous performance monitoring, to ensure the system operates at peak efficiency and reliability. When a minor issue is detected through the monitoring system, the facility's trained maintenance staff quickly diagnoses and resolves the problem, minimizing any disruption to the hospital's operations.

By prioritizing proper installation, maintenance, and monitoring of your commercial geothermal system, you can ensure its long-term performance, efficiency, and reliability while maximizing the benefits of this clean, sustainable energy source. Remember that investing in quality installation, regular maintenance, and ongoing monitoring can help you avoid costly repairs, minimize downtime, and optimize your geothermal system's contributions to your business's energy needs and sustainability goals.

As you embark on your journey to harness geothermal energy for your business, keep in mind the importance of careful planning, design, installation, and ongoing management of your geothermal system. By working with experienced professionals and dedicating the necessary resources to these critical aspects, you can unlock the full potential of geothermal energy to power your business while reducing your environmental impact and operating costs.

Chapter 11
Designing a Geothermal System for Your Business
Evaluating Your Business's Geothermal Potential

Geothermal energy is a clean, reliable, and efficient renewable energy source that can provide significant benefits for businesses looking to reduce their carbon footprint and energy costs. Designing a geothermal system for your business involves several key steps, including evaluating your geothermal potential, choosing the right system type and size, and planning for installation and maintenance. In this chapter, we'll start by focusing on evaluating your business's geothermal potential.

Evaluating Your Business's Geothermal Potential
Before investing in a geothermal system for your business, it's essential to assess whether your location has suitable geothermal resources and whether a geothermal system is a feasible and cost-effective solution for your energy needs. Evaluating your business's geothermal potential involves considering factors such as geology, land availability, and energy requirements. Here's a step-by-step guide to evaluating your business's geothermal potential:

Step 1: Assess Geological Conditions
The first step in evaluating your geothermal potential is to assess the geological conditions at your location. Geothermal resources are most accessible in areas with high heat flow, such as near tectonic plate boundaries, volcanic regions, or deep sedimentary basins. Factors to consider include:

- Geothermal gradient: The geothermal gradient is the rate at which temperature increases with depth in the Earth's crust. Areas with higher geothermal gradients are more suitable for geothermal energy development.
- Rock type and permeability: The type and permeability of the rock formations at your location can impact the feasibility and cost of drilling geothermal wells and extracting heat.
- Groundwater availability: The presence of groundwater can enhance heat transfer and improve the efficiency of geothermal systems. Assess the depth, temperature, and flow rate of any aquifers at your location.

To gather this information, consult local geological maps, geothermal resource assessments, and geothermal energy experts who can provide site-specific evaluations.

Step 2: Evaluate Land Availability and Accessibility
Next, evaluate the land availability and accessibility at your business location. Geothermal systems require sufficient space for drilling wells, installing heat exchangers, and constructing any necessary surface facilities. Consider the following factors:

- Land ownership: Ensure that you have the necessary rights and permissions to develop geothermal resources on your property or any adjacent land.
- Drilling access: Assess the accessibility of your site for drilling rigs and other equipment needed to construct geothermal wells. Consider any logistical challenges, such as steep terrain or limited road access.
- Surface facilities: Determine the space requirements for any surface facilities, such as heat exchangers, pumps, and piping, and ensure that you have adequate room for their installation and maintenance.

Step 3: Determine Energy Requirements and Efficiency
To properly size your geothermal system, you'll need to determine your business's energy requirements and the efficiency of the proposed system. This involves:

- Energy audit: Conduct a comprehensive energy audit to assess your current energy consumption, including heating, cooling, and electricity needs. This will help you determine the required capacity of your geothermal system.
- System efficiency: Evaluate the efficiency of different geothermal system types and configurations to determine which option best meets your energy needs and maximizes the use of your geothermal resources. Factors to consider include heat exchanger design, pumping requirements, and distribution systems.
- Cost-benefit analysis: Perform a cost-benefit analysis to compare the upfront costs and long-term savings of a geothermal system against conventional energy sources.

- Consider factors such as installation costs, operating and maintenance expenses, and potential energy savings over the system's lifespan.

Step 4: Consult with Geothermal Professionals
To ensure a thorough and accurate evaluation of your geothermal potential, consult with experienced geothermal professionals, such as geologists, engineers, and system designers. These experts can provide valuable insights and recommendations based on your specific location, energy needs, and business goals. They can also assist with site assessments, system design, and cost estimations.

Imagine a manufacturing facility located near a known geothermal resource area. The facility conducts a thorough geological assessment and determines that the site has a high geothermal gradient and suitable rock formations for drilling. They also evaluate their land availability and find sufficient space for the necessary wells and surface facilities. After conducting an energy audit and cost-benefit analysis, they determine that a geothermal system can provide a significant portion of their heating and cooling needs while reducing their long-term energy costs. The facility then engages a team of geothermal professionals to develop a detailed system design and implementation plan.

By carefully evaluating your business's geothermal potential, you can make an informed decision about whether a geothermal system is a practical and cost-effective solution for your energy needs. Keep in mind that a thorough assessment is crucial for the success and long-term viability of your geothermal project.

Choosing the Right Commercial Geothermal System

Once you have evaluated your business's geothermal potential and determined that a geothermal system is a viable option, the next step is to choose the right system for your commercial needs. Selecting the appropriate geothermal system involves considering factors such as system type, size, efficiency, and compatibility with your existing infrastructure. In this section, we'll discuss how to choose the right commercial geothermal system for your business.

1. Determine System Type
There are three main types of commercial geothermal systems: direct use, ground source heat pumps (GSHPs), and enhanced geothermal systems (EGS). Consider the following factors when determining the most suitable system type for your business:

- Direct use systems: These systems use geothermal resources directly for heating or cooling applications, such as space heating, water heating, or industrial processes. Direct use systems are most suitable for businesses with access to high-temperature geothermal resources and a consistent demand for heat.
- Ground source heat pumps (GSHPs): GSHPs use the relatively constant temperature of the Earth's shallow subsurface to provide heating and cooling. They are the most common type of commercial geothermal system and are suitable for a wide range of businesses, particularly those with balanced heating and cooling needs.

- Enhanced geothermal systems (EGS): EGS involve creating an artificial geothermal reservoir by fracturing hot rock formations and circulating water through the created fractures. These systems are most suitable for businesses with access to deep, high-temperature geothermal resources but limited natural permeability.

2. Evaluate System Size and Capacity

The size and capacity of your geothermal system will depend on your business's energy requirements, as determined by your energy audit. When evaluating system size and capacity, consider the following factors:

- Peak load: Determine the peak heating and cooling load of your business to ensure that the geothermal system can meet your maximum energy demand.
- Annual energy consumption: Calculate your annual energy consumption for heating and cooling to properly size your geothermal system for optimal performance and efficiency.
- Load profile: Analyze your business's daily and seasonal load profile to identify any variations in energy demand that may impact system sizing and design.
- Redundancy and backup: Consider incorporating redundancy or backup systems to ensure uninterrupted operation in case of maintenance or equipment failure.

3. Assess System Efficiency and Performance

When choosing a commercial geothermal system, assess the efficiency and performance of different options to maximize energy savings and minimize operating costs. Key factors to consider include:

- Coefficient of Performance (COP): The COP is a measure of a geothermal system's efficiency in providing heating or cooling. Higher COPs indicate more efficient systems that can deliver more energy output per unit of input.
- Energy Efficiency Ratio (EER): The EER is a measure of a geothermal system's cooling efficiency, calculated as the ratio of cooling capacity to power input. Higher EERs indicate more efficient cooling performance.
- Heat exchanger effectiveness: The effectiveness of the heat exchanger in transferring heat between the geothermal fluid and the building's heating and cooling system can significantly impact overall system efficiency.
- Pumping requirements: Evaluate the pumping requirements for circulating geothermal fluid and consider the associated energy consumption and operating costs.

4. Consider Integration and Compatibility

When selecting a commercial geothermal system, consider its integration and compatibility with your existing building infrastructure and energy systems. Factors to evaluate include:

- HVAC system compatibility: Ensure that the geothermal system is compatible with your existing heating, ventilation, and air conditioning (HVAC) equipment, such as air handlers, ductwork, and control systems.
- Electrical system compatibility: Verify that your building's electrical system can accommodate the power requirements of the geothermal system, including any necessary upgrades or modifications.

- Spatial requirements: Consider the spatial requirements for the geothermal system components, such as heat exchangers, pumps, and piping, and ensure that your facility has adequate space for their installation and maintenance.

5. *Evaluate Costs and Return on Investment*
Finally, evaluate the costs and return on investment (ROI) of different commercial geothermal system options to determine the most cost-effective solution for your business. Consider the following factors:

- Initial installation costs: Obtain detailed cost estimates for the installation of the geothermal system, including well drilling, heat exchangers, pumps, piping, and any necessary building modifications.
- Operating and maintenance costs: Estimate the ongoing operating and maintenance costs of the geothermal system, including electricity consumption, regular maintenance, and any potential repairs or replacements.
- Energy savings and ROI: Calculate the projected energy savings and ROI of the geothermal system based on your current energy costs, system efficiency, and estimated lifespan. Compare these savings to the upfront installation and ongoing operating costs to determine the financial viability of the project.

Imagine a large office complex that has determined that a ground source heat pump system is the most suitable geothermal option for their energy needs. They work with a geothermal system designer to properly size the system based on their peak load, annual energy consumption, and load profile. The designer recommends a high-efficiency heat pump with a COP of 4.5 and an EER of 20, coupled with a closed-loop ground heat exchanger. The system is designed to integrate seamlessly with the building's existing HVAC infrastructure and electrical system. After evaluating the installation and operating costs, the office complex determines that the geothermal system will provide a significant return on investment through reduced energy costs and lower maintenance requirements over its 25-year lifespan.

By carefully choosing the right commercial geothermal system based on your business's specific energy requirements, geothermal resources, and budget, you can maximize the benefits of this clean, reliable, and efficient renewable energy source. Keep in mind that proper system selection, sizing, and design are critical for the long-term performance, cost-effectiveness, and sustainability of your geothermal investment.

In the next section, we'll discuss the installation, maintenance, and monitoring considerations for commercial geothermal systems to ensure optimal performance and longevity.

Installation, Maintenance, and Monitoring

After selecting the right commercial geothermal system for your business, the next crucial steps are proper installation, regular maintenance, and ongoing monitoring to ensure optimal performance, efficiency, and longevity. Installing a geothermal system involves several key considerations, such as site preparation, drilling, and system integration, while regular maintenance and monitoring are essential for identifying and addressing any issues that may arise. In this section, we'll discuss the installation, maintenance, and monitoring requirements for your commercial geothermal system.

Installation

1. Site Preparation and Permitting

Before installing your geothermal system, properly prepare the site and obtain all necessary permits and approvals. This includes:

- Conducting a detailed site survey to identify any potential obstacles or constraints, such as underground utilities or environmental sensitivities.
- Obtaining all required permits and approvals from local, state, and federal authorities, such as building permits, drilling permits, and environmental assessments.
- Preparing the site for drilling and installation, including clearing and grading the area, establishing access roads, and setting up any necessary safety measures.

2. Drilling and Well Installation

The drilling and installation of geothermal wells is a critical component of the installation process. This involves:

- Drilling the geothermal wells to the required depth and diameter based on the system design and geological conditions. This may involve using specialized drilling rigs and techniques, such as directional drilling or casing installation.
- Installing the well casing, grouting, and any necessary well screens or packers to ensure the integrity and longevity of the well.
- Conducting well logging and testing to verify the well's integrity, productivity, and thermal properties.

3. Heat Exchanger and Piping Installation

After the wells are installed, the heat exchanger and piping system must be installed to transfer heat between the geothermal fluid and the building's HVAC system. This includes:

- Installing the heat exchanger, circulation pumps, and any necessary valves or controls based on the system design and manufacturer's specifications.
- Connecting the heat exchanger to the geothermal well piping and the building's HVAC system, ensuring proper insulation and sealing to minimize heat loss.
- Pressure testing and flushing the piping system to verify its integrity and remove any debris or contaminants.

4. System Integration and Commissioning

The final step in the installation process is integrating the geothermal system with the building's existing HVAC and electrical systems and commissioning the system for operation. This involves:

- Connecting the geothermal system to the building's HVAC system, including any necessary modifications or upgrades to the existing ductwork, piping, or control systems.
- Installing and programming the system controls, monitoring equipment, and any necessary safety devices or alarms.
- Conducting a thorough commissioning process to verify that the system is operating as designed, including testing and balancing the system, calibrating sensors and controls, and providing operator training.

Maintenance and Monitoring

1. Regular Inspection and Testing

To ensure the ongoing performance and efficiency of your commercial geothermal system, conduct regular inspections and testing, including:

- Visual inspections of the wells, piping, heat exchanger, and other system components to identify any signs of wear, damage, or leakage.
- Pressure and flow testing to verify that the system is operating within the designed parameters and to detect any blockages or restrictions.

- Water quality testing to monitor the geothermal fluid's chemistry and to identify any potential scaling, corrosion, or biological growth issues.

2. Preventive Maintenance

Implement a comprehensive preventive maintenance program to proactively address any potential issues and extend the lifespan of your geothermal system. This may include:

- Regularly cleaning and flushing the heat exchanger and piping system to remove any accumulated debris or scale.
- Lubricating and replacing any worn or damaged components, such as pumps, valves, or bearings, based on the manufacturer's recommendations.
- Updating and calibrating the system controls and monitoring equipment to ensure optimal performance and efficiency.

3. Performance Monitoring and Optimization

Continuously monitor and optimize your geothermal system's performance to maximize energy savings and minimize operating costs. This involves:

- Installing and utilizing a comprehensive monitoring system to track key performance indicators, such as energy consumption, system efficiency, and operating temperatures.
- Analyzing performance data to identify any trends, anomalies, or opportunities for improvement, such as adjusting system setpoints or reconfiguring piping layouts.

- Implementing any necessary system modifications or upgrades to optimize performance and efficiency based on the monitoring data and analysis.

4. Emergency Response and Troubleshooting

Establish a clear protocol for responding to any system emergencies or malfunctions and troubleshooting any issues that arise. This includes:

- Developing a detailed emergency response plan that outlines the procedures for shutting down the system, isolating any damaged components, and notifying relevant personnel.
- Training facility staff and maintenance personnel on the proper operation and troubleshooting of the geothermal system, including identifying and diagnosing common issues.
- Establishing a network of qualified geothermal service providers and equipment suppliers to quickly respond to any emergencies or repair needs.

Imagine a hospital that has installed a large-scale ground source heat pump system to provide heating and cooling for its facilities. The installation process involved careful site preparation, including a detailed site survey and obtaining all necessary permits and approvals. The geothermal wells were drilled to a depth of 500 feet using specialized directional drilling techniques, and the well casings were installed and grouted to ensure long-term integrity.

The heat exchanger and piping system were installed and integrated with the hospital's existing HVAC system, and the entire system was thoroughly commissioned to verify optimal performance.

The hospital implements a comprehensive maintenance and monitoring program, including regular inspections, preventive maintenance, and continuous performance monitoring, to ensure the system operates at peak efficiency and reliability. When a minor issue is detected through the monitoring system, the facility's trained maintenance staff quickly diagnoses and resolves the problem, minimizing any disruption to the hospital's operations.

By prioritizing proper installation, maintenance, and monitoring of your commercial geothermal system, you can ensure its long-term performance, efficiency, and reliability while maximizing the benefits of this clean, sustainable energy source. Remember that investing in quality installation, regular maintenance, and ongoing monitoring can help you avoid costly repairs, minimize downtime, and optimize your geothermal system's contributions to your business's energy needs and sustainability goals.

As you embark on your journey to harness geothermal energy for your business, keep in mind the importance of careful planning, design, installation, and ongoing management of your geothermal system. By working with experienced professionals and dedicating the necessary resources to these critical aspects, you can unlock the full potential of geothermal energy to power your business while reducing your environmental impact and operating costs.

Part IV: Financing and Incentives for Renewable Energy
Chapter 12
Government Incentives and Tax Credits

Investing in renewable energy systems can be a significant financial decision for both homeowners and businesses. Fortunately, many governments around the world offer various incentives and tax credits to encourage the adoption of clean energy technologies. These incentives can help offset the upfront costs of installing renewable energy systems, making them more accessible and cost-effective for a wider range of consumers. In this chapter, we'll explore the different types of government incentives and tax credits available for renewable energy projects.

1. Federal Tax Credits

Federal tax credits are one of the most significant incentives for renewable energy in many countries. In the United States, for example, the Investment Tax Credit (ITC) and the Production Tax Credit (PTC) are two key federal incentives for renewable energy.

- Investment Tax Credit (ITC): The ITC is a tax credit that allows homeowners and businesses to deduct a percentage of the cost of installing a renewable energy system from their federal income taxes.

- As of 2021, the ITC provides a 26% tax credit for solar, geothermal, and small wind systems, with the credit scheduled to step down to 22% in 2023 and 10% for commercial projects in 2024 and beyond.
- Production Tax Credit (PTC): The PTC is a tax credit based on the amount of electricity generated by a renewable energy system. It primarily benefits large-scale wind and geothermal projects, providing a per-kilowatt-hour tax credit for the first ten years of a system's operation.

2. State and Local Incentives

In addition to federal incentives, many states and local governments offer their own incentives for renewable energy projects. These can include:

- Rebates: Some states and utilities offer rebates for installing renewable energy systems, which can significantly reduce the upfront cost of installation. Rebates may be based on the size of the system, its expected energy production, or other factors.
- Grants: Grants are another form of financial assistance that can help cover the cost of installing renewable energy systems. Grants may be available from state or local governments, utilities, or non-profit organizations.
- Tax Credits and Exemptions: Many states offer their own tax credits or exemptions for renewable energy projects, which can be claimed in addition to federal tax credits. These may include property tax exemptions, sales tax exemptions, or income tax credits.

- Performance-Based Incentives (PBIs): PBIs are incentives that pay system owners based on the actual energy production of their renewable energy system. These can include feed-in tariffs, which guarantee a fixed price for the electricity generated, or renewable energy certificates (RECs), which represent the environmental attributes of renewable energy generation.

3. Net Metering and Virtual Net Metering

Net metering and virtual net metering are policies that allow renewable energy system owners to offset their electricity bills by sending excess energy back to the grid.

- Net Metering: Under a net metering policy, when a renewable energy system generates more electricity than the owner consumes, the excess energy is sent back to the grid, spinning the owner's electricity meter backward. This allows the owner to offset their electricity consumption and potentially earn credits for future energy use.
- Virtual Net Metering: Virtual net metering extends the benefits of net metering to multi-tenant buildings or community solar projects. It allows multiple customers to share the benefits of a single renewable energy system, with each customer receiving a portion of the energy credits based on their share of the system.

4. Accelerated Depreciation

Accelerated depreciation is a tax incentive that allows businesses to depreciate the cost of a renewable energy system more quickly than traditional depreciation schedules.

This can provide significant tax benefits and improve the financial viability of renewable energy projects for businesses.

5. Loan Programs and Financing Options

Many governments and financial institutions offer special loan programs or financing options for renewable energy projects. These can include:

- Low-Interest Loans: Government-backed low-interest loans can help homeowners and businesses finance renewable energy projects with more favorable terms than traditional loans.
- Property Assessed Clean Energy (PACE) Financing: PACE financing allows property owners to borrow money for renewable energy projects and repay the loan through their property tax bills. This can provide long-term, low-interest financing that stays with the property, even if it is sold.
- Green Bonds: Green bonds are debt securities issued by governments, companies, or financial institutions to finance environmentally friendly projects, including renewable energy projects. These bonds can provide a source of low-cost capital for large-scale renewable energy development.

Imagine a small business owner who is considering installing a rooftop solar system to reduce their electricity costs and environmental impact. By taking advantage of the federal Investment Tax Credit, state rebates, and a low-interest loan program, the business owner is able to offset a significant portion of the upfront cost of the solar installation.

Additionally, the business is able to take advantage of accelerated depreciation to further reduce their tax liability. With these incentives and financing options, the solar project becomes a financially viable and attractive investment for the small business.

Government incentives and tax credits can play a crucial role in making renewable energy projects more accessible and cost-effective for homeowners and businesses. By understanding the various incentives available at the federal, state, and local levels, as well as exploring financing options like low-interest loans and PACE financing, individuals and organizations can better navigate the financial landscape of renewable energy adoption.

In the next chapter, we'll delve deeper into the financing options available for renewable energy projects, including traditional loans, leasing, and power purchase agreements, to help you identify the best financing strategy for your specific needs and goals.

Chapter 13
Financing Options for Renewable Energy Projects

While government incentives and tax credits can significantly reduce the upfront costs of renewable energy projects, many homeowners and businesses still require financing to make these projects a reality. Fortunately, there are several financing options available that cater to the unique needs and circumstances of different individuals and organizations. In this chapter, we'll explore the various financing options for renewable energy projects, including traditional loans, leasing, and power purchase agreements.

1. Cash Purchase
For those with sufficient capital, a cash purchase is the most straightforward way to finance a renewable energy project. By paying for the system outright, the owner can avoid interest payments and immediately begin benefiting from reduced energy costs and any available incentives. However, this option may not be feasible for everyone due to the high upfront costs of renewable energy systems.

2. Traditional Loans
Traditional loans, such as home equity loans or business loans, can be used to finance renewable energy projects. These loans typically offer competitive interest rates and allow the borrower to own the system outright. However, they may require collateral and a strong credit score to qualify.

- Home Equity Loans: Homeowners can leverage the equity in their homes to secure a loan for a renewable energy project. This type of loan often offers lower interest rates than unsecured loans, as the home serves as collateral.
- Business Loans: Businesses can seek traditional business loans from banks or other financial institutions to finance renewable energy projects. These loans may be secured by business assets or require a personal guarantee from the business owner.

3. Leasing

Leasing is an increasingly popular option for financing renewable energy projects, particularly for solar installations. Under a lease agreement, a third-party owner installs and maintains the renewable energy system on the customer's property, while the customer makes monthly lease payments. This allows the customer to benefit from reduced energy costs without the high upfront costs of ownership. There are two main types of leases:

- Capital Lease: In a capital lease, the customer is considered the owner of the system for tax purposes and can take advantage of any available incentives and depreciation benefits. At the end of the lease term, the customer typically has the option to purchase the system at a reduced price.
- Operating Lease: In an operating lease, the third-party owner maintains ownership of the system and receives any available incentives and tax benefits. The customer simply makes monthly lease payments and benefits from reduced energy costs.

At the end of the lease term, the customer may have the option to renew the lease, purchase the system, or have it removed.

4. Power Purchase Agreements (PPAs)

Power Purchase Agreements are similar to leases, but instead of making fixed monthly payments, the customer agrees to purchase the electricity generated by the renewable energy system at a predetermined price per kilowatt-hour. The third-party owner installs, owns, and maintains the system, while the customer benefits from reduced energy costs and a predictable electricity rate. PPAs are commonly used for larger-scale projects, such as commercial solar installations or community solar projects.

5. Property Assessed Clean Energy (PACE) Financing

PACE financing, as mentioned in the previous chapter, allows property owners to finance renewable energy projects through their property tax bills. The loan is attached to the property rather than the individual, and repayments are made through an additional assessment on the property tax bill. This type of financing can offer long repayment terms and potentially lower interest rates, making it an attractive option for many property owners.

6. Green Bonds and Crowdfunding

For larger-scale renewable energy projects, green bonds and crowdfunding can provide alternative financing options.

- Green Bonds: As discussed earlier, green bonds are debt securities issued to finance environmentally friendly projects, including renewable energy projects.

- These bonds can be issued by governments, companies, or financial institutions and can provide a source of low-cost capital for large-scale projects.
- Crowdfunding: Crowdfunding platforms allow renewable energy projects to raise capital from a large number of individual investors. This can be particularly useful for community-based projects or innovative technologies that may struggle to secure traditional financing.

Imagine a homeowner who wants to install a rooftop solar system but doesn't have the cash to purchase the system outright. After exploring their options, the homeowner decides to pursue a solar lease. The leasing company installs and maintains the solar panels on the homeowner's roof, while the homeowner makes fixed monthly lease payments that are lower than their previous electricity bills. This allows the homeowner to benefit from clean energy and reduced utility costs without the high upfront investment or maintenance responsibilities.

Financing options for renewable energy projects have evolved significantly in recent years, providing a range of solutions for different needs and circumstances. By understanding the various financing options available, from traditional loans to innovative models like PACE financing and crowdfunding, homeowners and businesses can identify the best approach for their specific renewable energy goals.

In the next chapter, we'll explore the long-term financial benefits of renewable energy projects, including return on investment and potential savings, to help you make informed decisions about your clean energy investments.

Chapter 14
Return on Investment and Long-Term Savings

Investing in renewable energy projects can provide significant long-term financial benefits, in addition to the environmental and social advantages of clean energy. By understanding the potential return on investment (ROI) and long-term savings associated with renewable energy systems, homeowners and businesses can make informed decisions about their clean energy investments. In this chapter, we'll explore the key factors that influence the ROI and long-term savings of renewable energy projects, as well as some tools and methods for calculating these benefits.

1. Factors Influencing ROI and Long-Term Savings
Several factors can impact the ROI and long-term savings of renewable energy projects:
- Initial Cost: The upfront cost of the renewable energy system, including equipment, installation, and any necessary upgrades or modifications, will significantly influence the ROI and payback period.
- Energy Production: The amount of energy generated by the renewable energy system will directly impact the long-term savings and ROI. Factors such as system size, efficiency, and location-specific conditions (e.g., solar irradiation or wind speeds) will affect energy production.
- Energy Prices: The price of electricity or fuel that the renewable energy system is offsetting will play a significant role in determining the long-term savings. Higher energy prices will generally lead to greater savings and a faster ROI.

- Incentives and Credits: Government incentives, tax credits, and other financial support can substantially reduce the initial cost of the renewable energy system, improving the ROI and shortening the payback period.
- System Lifespan: The expected lifespan of the renewable energy system will impact the total long-term savings. Many renewable energy technologies, such as solar panels and geothermal systems, have expected lifespans of 25 years or more, allowing for significant cumulative savings over time.

2. Calculating ROI and Payback Period

To calculate the ROI and payback period for a renewable energy project, you'll need to consider the following factors:

- Net Cost: Determine the net cost of the renewable energy system by subtracting any incentives, rebates, or tax credits from the initial cost.
- Annual Energy Savings: Estimate the annual energy savings by multiplying the expected energy production by the current price of electricity or fuel. Be sure to account for any expected increases in energy prices over time.
- Annual Maintenance Costs: Factor in any annual maintenance costs associated with the renewable energy system, such as cleaning, inspections, or repairs.
- Payback Period: Calculate the payback period by dividing the net cost by the annual energy savings minus the annual maintenance costs. This will give you the number of years it will take for the cumulative savings to equal the initial investment.

- ROI: To calculate the ROI, divide the total lifetime savings (annual energy savings multiplied by the system lifespan) by the net cost, and express the result as a percentage.

3. Online Tools and Calculators

Several online tools and calculators are available to help estimate the ROI and long-term savings of renewable energy projects:

- PVWatts: Developed by the National Renewable Energy Laboratory (NREL), PVWatts is a web-based tool that estimates the energy production and cost savings of grid-connected solar PV systems.
- RETScreen: RETScreen is a comprehensive clean energy management software platform that helps assess the feasibility and performance of renewable energy, energy efficiency, and cogeneration projects.
- Wind Energy Payback Period Worksheet: Developed by the U.S. Department of Energy, this worksheet helps estimate the payback period and cost savings for small wind electric systems.
- Geothermal Electricity Technology Evaluation Model (GETEM): Developed by the U.S. Department of Energy, GETEM is a computer model for evaluating the economic performance of geothermal power projects.

4. Long-Term Savings and Energy Independence

Beyond the direct financial benefits, investing in renewable energy projects can provide long-term savings and increased energy independence for homeowners and businesses.

- Protection Against Energy Price Volatility: By generating their own clean energy, consumers can protect themselves from potential increases or fluctuations in energy prices over time.
- Increased Property Value: Studies have shown that homes and commercial properties with renewable energy systems often have higher property values and sell faster than comparable properties without these systems.
- Energy Independence: Renewable energy projects can help reduce dependence on the grid and provide a level of energy independence, particularly when combined with energy storage solutions.

Imagine a small business that invests in a 50 kW rooftop solar PV system. The total cost of the system is $100,000, but with a 26% federal tax credit and a $10,000 state rebate, the net cost is reduced to $64,000. The system is expected to generate 65,000 kWh of electricity annually, offsetting $9,750 in utility costs (assuming an electricity price of $0.15/kWh). With an expected lifespan of 25 years and annual maintenance costs of $500, the payback period is calculated to be approximately 6.8 years, and the ROI is estimated at 152%. Over the 25-year lifespan of the system, the business is projected to save over $243,000 in energy costs, providing significant long-term financial benefits.

Understanding the potential ROI and long-term savings of renewable energy projects is crucial for making informed investment decisions. By considering factors such as initial costs, energy production, incentives, and system lifespan, and utilizing available tools and calculators, homeowners and businesses can assess the financial viability of their clean energy investments.

Conclusion

Throughout this book, we have explored the incredible potential of renewable energy technologies to transform the way we power our homes and businesses. By harnessing the clean, abundant energy sources that surround us – the sun, the wind, and the Earth itself – we can create a more sustainable, resilient, and prosperous future for ourselves and generations to come.

As we have seen, each renewable energy technology – solar, wind, and geothermal – offers unique benefits and challenges, and the key to success lies in carefully evaluating your specific needs, resources, and goals to design a system that maximizes both performance and return on investment. Whether you are a homeowner looking to reduce your carbon footprint and energy bills, or a business owner seeking to boost your bottom line and demonstrate your commitment to sustainability, renewable energy offers a powerful solution.

But the benefits of renewable energy extend far beyond individual homes and businesses. By accelerating the transition to a clean energy economy, we can combat the urgent threat of climate change, improve public health by reducing air and water pollution, and create millions of new jobs in the growing renewable energy sector. We can also enhance our energy security and resilience by diversifying our energy mix and reducing our dependence on finite, volatile fossil fuels.

Of course, the path to a sustainable energy future is not always easy, and it requires a collective effort from policymakers, industry leaders, and citizens alike. We must continue to invest in research and development to improve the efficiency and affordability of renewable energy technologies, and we must work to overcome the barriers – whether technical, financial, or political – that can slow their adoption.

But as the success stories and case studies throughout this book have shown, the renewable energy revolution is already well underway, and its momentum is only growing. From small-scale residential installations to large-scale commercial projects, people around the world are embracing the power of renewable energy to build a better future for themselves and for the planet.

So let this book be your guide and your inspiration as you embark on your own renewable energy journey. Whether you are taking your first steps or already well on your way, remember that every action, no matter how small, can make a difference. By choosing to harness the clean, abundant energy that surrounds us, you are not only investing in your own future, but also in the future of our shared world.

The road to a sustainable energy future may be long, but with determination, innovation, and collaboration, we can get there together. So let us move forward with confidence and purpose, knowing that the power to create a cleaner, greener, and more prosperous world is within our grasp. The future is bright, and it is powered by renewable energy.

www.ingramcontent.com/pod-product-compliance
Lightning Source LLC
Chambersburg PA
CBHW051534240526
45471CB00020B/1787